Quality Assurance in Adhesive Technology

Quality Assurance in Adhesive Technology

EUREKA Project EU716

ABINGTON PUBLISHING
Woodhead Publishing Limited in association with The Welding Institute
Cambridge England

EUREKA Project EU716, Centre for Adhesive Technology, TWI and School of Industrial & Manufacturing Science, Cranfield University

UK industrial participants – British Steel Technical, Carello Lighting (now Magneti Marelli), CarnaudMetalbox, Ciba Polymers, Commercial Hydraulics Keelavite, Hunting Engineering, Permabond, Pilkington, Westland Aerospace

Financial support – Department of Trade and Industry, UK

European partners – Institute of Production Engineering Research (IVF), Goteborg, Sweden; Adhesion Institute, Delft, Netherlands

Document authors – Dr Alan W Espie, Centre of Adhesive Technology, TWI; Professor John H Rogerson and Dr Kambiz Ebtehaj, School of Industrial & Manufacturing Science, Cranfield University

Software author – Dr Kambiz Ebtehaj

Published by Abington Publishing
Abington Hall, Abington
Cambridge CB1 6AH, England

First published 1998, Abington Publishing

British Library Cataloguing in Publication Data
A catalogue record for this book is available from the British Library.

ISBN 1 85573 259 9

Printed by Antony Rowe, Wiltshire, England.

CONTENTS

APPENDIX A MANUFACTURING PROCESS CHECKLISTS

Adhesion Institute, Delft	GRP composite pipework
Anonymous	Flexible packaging lamination
British Steel Technical	Laminated floor panel
Carello Lighting	Headlamp assembly
Commercial Hydraulics Keelavite	Power unit structural framework
Hunting Engineering	Law 80 launch tube
IBM UK	Disk drive actuator
Oxford Brookes University	Construction joint sealing
Permabond	Needle electrode
Pilkington	Automotive glazing
Westland Aerospace	Aircraft wing vane
Blank checklist	

APPENDIX B CHECKLIST DOCUMENTS

Specification review checklists	Quality requirements
	Concerned parties and data resources
	Specification change
Design checklists	Design
	Adherend materials
	Consumable materials
	Adhesive
	Adhesive storage
	Design of process

Quality Assurance in Adhesive Technology (EU716)

EXECUTIVE SUMMARY

Background

A three year project finished in March 1995, called Quality Assurance in Adhesive Technology (EUREKA Project, EU716, QUASIAT). It involved the combined activities of the Centre for Adhesive Technology at TWI, Cambridge, Cranfield University, a group of nine UK industrial participants and cooperative work in the Netherlands and Sweden. The project was funded in the UK by industry and the DTI.

Objective

The objective of the work has been to obtain improved reliability of adhesively bonded products and with this specific package of help, to assist the application of general quality management systems already in place in manufacturing industry, to the specific issue of adhesive bonding.

Adhesive bonding can be considered as a "special process", the results of which cannot be fully verified by subsequent inspections and testing. Therefore continuous monitoring and compliance with documented procedures are required to provide assurance of quality.

Conclusions

When QUASIAT was planned, the emphasis was intended to be on the production phase, but as the work progressed, it became more and more apparent that it must emphasise the need to include quality considerations in the design stage (specifying materials, adhesive and process) and not only when production actually begins.

A generic quality management model has been developed by Cranfield University, in both text and software formats, which considers all the major stages from design through to final assembly and inspection, following all the steps of the process and their control points, which can be made to contribute to reliable assembly. It finally delivers a quality plan for a specific bonding application. Please see page 62 for conditions of use of the software.

Worked examples of quality tools and techniques have been developed with the participating companies, for several of the techniques considered to be of direct value in adhesive technology. By exemplifying their use in relevant product assembly using adhesives, we hope to dispel the misconceptions that can prevent acceptance.

Recommendations

With the assistance of the outputs of QUASIAT, the companies involved in the project and others will be able to achieve higher levels of reliability in their adhesive bonding processes. In addition, manufacturers new to the technology can adopt adhesives with greater confidence.

1 INTRODUCTION TO QUASIAT

A three year collaborative EUREKA project, **QU**ality **AS**surance **I**n **A**dhesive **T**echnology [**QUASIAT**, EU716] started in March 1992. The objectives were to obtain improved reliability of adhesively bonded products and with this specific help package, to aid the more general quality management systems already in place in manufacturing industry. It involved the combined activities of the Centre for Adhesive Technology at TWI, Cambridge, the School of Industrial and Manufacturing Science at Cranfield University, a group of nine UK industrial participants [British Steel Technical, Carello Lighting, CarnaudMetalbox, Ciba Polymers, Commercial Hydraulics Keelavite, Hunting Engineering, Permabond, Pilkington and Westland Aerospace] plus cooperative work at the Adhesion Institute, Delft, Netherlands and the Institute for Production Engineering and Research [IVF], Göteborg, Sweden. It was funded in the UK by the above industries and the Department of Trade and Industry [DTI].

Adhesives are used today in almost all sections of industry, from aircraft components to convenience food packaging. They have traditionally been seen as important if they carry out a load-bearing or high integrity function, but adhesives are now a vital component of many modern products. All the industries who participated in QUASIAT considered their products to be `design-critical', whether the product was a thrust reverser for an aero engine, a sealed-beam headlamp unit, or flexible laminated packaging to contain a freezer-to-microwave meal. In the past the adhesive joint has been too often considered a low-cost, rather than a high-value and special process in product assembly.

By far the most difficult task facing the engineer in the design and manufacture of an adhesively bonded product is to ensure that an acceptable level of reliability and consistency is achieved in the joint. Inspection of the final product, which has long been employed as an assurance for the quality of adhesive bonding, is expensive, is not always applicable and is 'after the event'. Reliability and consistency have to be achieved by thorough in-process control from the time the raw materials come in until the last finishing operation is completed. Adhesive bonding can be a more complex procedure to control than other joining methods. For example, on a car assembly line spot welding is the responsibility of one station, but up to five points of the line can contribute to success or failure of a bonded joint.

Companies committed to total quality management aim to practice defect prevention, not detection, and to have stable controlled processes. They do this to ensure they meet their customers' needs and expectations, and to improve their own efficiency and profitability. Quality systems as defined by the ISO 9000 series give evidence that a manufacturer is committed to quality improvement. However, they do not in themselves guarantee a high quality product as the standard assumes that the **product** quality is separately defined. Enhanced product quality will have a direct influence on both product reliability and liability. In terms of the ISO 9000 series, adhesive bonding must be considered as a special process where particular care and control are required, since in most cases the results cannot be fully verified by subsequent inspections and testing. Therefore continuous monitoring and control of process parameters, or qualified operators and/or compliance with documented procedures are required.

Work started with the analysis of current practice and problems in the quality management of adhesive bonding. The design stages of both product and assembly were also scrutinised, so that the implications of design on quality management, or vice versa, could be determined. When QUASIAT was planned, the focus was to be on the production phase, but as the work progressed, it became more apparent that it must emphasise the need to include quality considerations at the design stage, specifying materials, adhesive and process, and thus specifying the manufacturing process instructions (or the 'post-design audit items'!). In many instances, the responsibility for quality management has been delegated exclusively to production functions. However, effort spent prior to production can provide significant cost benefits, such as ensuring that all problems have been resolved before volume production. Therefore the output of this work is targeted towards those in product development and design as well as manufacture and assembly.

A generic quality management model has been developed by Cranfield University, which considers all the major stages from design through to final assembly and inspection, following all the steps of the process and their control points, which can be made to contribute to reliable assembly. This model was developed as the result of detailed analyses of several bonding applications provided by the industrial participants in the project. The analysis was conducted via a detailed questionnaire plus subsequent visits and discussions to establish the details of each application. The model is designed to accommodate a wide range of applications, but at the same time it can be used to provide a detailed quality plan for an individual assembly process.

Many quality tools and techniques have been developed to help manufacturing industry, almost all defined by acronyms, such as FMEA, QFD or SPC. Worked examples have been collected from the participating companies, of techniques considered to be of direct value in adhesive technology. By exemplifying their use in relevant product assembly using adhesives, we hope to dispel the misconceptions that can prevent acceptance.

The ultimate objective is to provide both appreciation and practical tools for design and production engineers, so that the products and ultimately their customers can enjoy the benefits of reliable adhesive bonding. This package will be equally relevant to the precise, measured assembly of load bearing joints in defence and aerospace components; to the rapid, cell-based assembly of automotive components; to continuous, high speed production lines for packaging film or electronic components; and to labour-intensive, on-site work such as pipe bonding or sealant application, where there may be no factory process control but a reliance on the training and trust given to the employee. With the assistance of the outputs from QUASIAT, the companies involved in the project and then others will be able to achieve higher levels of reliability in their adhesive bonding processes. In addition, potential users of this technology will be more confident to adopt adhesive bonding.

A bibliography of selected further reading on quality management and adhesive technology is included at the end of this report.

European Interactions

EUREKA projects are funded by each National Government; therefore no money changes hands internationally and there is no common funding agency, as in some other European Union programmes. Collaboration is based on goodwill and cooperation towards a common objective and in this project the exchange of ideas and information between partners was most valuable.

The Dutch Adhesion Institute participated in a project which focused on the adhesive bonding of glass-reinforced composite pipework; the instructions for operatives, defining acceptable defect levels, developing and tuning suitable non-destructive test methods. This single application, which was studied in great detail, has some elements common to the structural bonding examples documented in UK factories but, since it is carried out on-site or off-shore, there are the similar problems of process control as seen in the UK example of sealant use in the construction industry.

In Sweden, the IVF had already carried out two years work prior to the start of QUASIAT, drafting the quality assurance chapter for an adhesive bonding handbook being developed in a collaborative project by the Nordic countries. Their continued work with Swedish manufacturing industry has led to additional manufacturing case studies, plus exemplification of modern quality tools and techniques applied to adhesive bonding.

Acknowledgements

This project was instigated by Professor A. Beevers, of Oxford Brookes University, while Chief Executive of the Centre for Adhesive Technology, along with Professor J. H. Rogerson, Cranfield University, plus industrialists including M. Hall, Xyratex, previously IBM UK, Havant.

P. Webster, a research student at Cranfield also contributed to the early part of the project.

Acknowledgement is given to Xyratex for permission to reproduce the paper on Taguchi techniques applied to adhesive bonding, by M. Hall and T. Twine.

In the early stages of the EUREKA project, J-P. Jeandrau, CETIM, St Etienne, France, also participated.

DTI project officers A. Roberts and latterly P. Hale are hereby thanked for their help and support.

2 QUALITY MANAGEMENT

2.1 General Introduction to Quality Management

What We Mean By Quality

Since quality as a business concept is so important then it is necessary, if only to ease communication and define requirements, to develop and agree a terminology and, consequent on this, develop a suitable quality infrastructure to support businesses (both private and public sector).

Before we do this, it is important to review the business issues which the term quality implies, from the point of view of a supplier of goods and/or services.

There are five separate but closely related issues to which the concept of quality can provide solutions:

1. Competitiveness.
2. Consumer concerns for safety.
3. Environmental protection.
4. Product liability.
5. Fitness for purpose.

Competitiveness

The aspect (or definition) of quality which is relevant is as follows:

Quality is the elimination of waste.

This broad concept which addresses the issue of how a company can improve its performance to outstrip its competitors, embraces such topics as:

> Control of purchasing of materials, components and services.
>
> Reduction of errors in manufacturing, handling and delivery.
>
> Efficient work handling, by adequate forethought during the design.
>
> Financial controls.

These indicate that a company must maximise the efficiency of its business processes by minimising waste of resources, be they materials, time, people, or money. If 'quality' is then seen as the elimination of waste, it becomes a significant management issue and not just a technical issue relating to product testing or the control of manufacturing tolerances.

Consumer concerns for safety

As part of increased consumer expectations, there is an increased emphasis on 'safety'. For example:

a. Safety features in automobiles (seat belts, air bags, side impact bars, anti-lock brakes).

b. More rigorous testing regimes for pharmaceuticals.

c. Better fire precautions in public buildings.

In all areas where 'safety' is an important feature, that is most products and many services, we are concerned with the safety of a 'system', e.g. a transportation system, such that risk of harm is reduced to an acceptable (although often arbitrary) level. Safety risk analysis is a complex subject with its links to reliability, the interaction of sub-systems and, not least, the identification of what is an acceptable risk.

However, in all cases, the safety of a 'system' will depend on the behaviour and consistency of behaviour of elements of hardware or processes. In other words, the quality of these elements is crucial.

'Safety' is a function of the design of a system. The system will behave as the design predicts if the components of the system (hardware and software) behave in a predictable manner. This means there must be assurance of the quality of these elements.

Quality, in this sense, has a more technological emphasis with the mechanism of managerial arrangements, process controls and product testing, ensuring two things:

a. The design is adequate (i.e. safe).

b. The manufacturing processes are under sufficient control to deliver the design requirements with minimum variability.

The link between this second objective and the issue of competitiveness by reduction of waste is clear, since minimum product variability is outside the customer's specification.

Environmental Protection

This is an analogous issue to that of safety. Public awareness has been raised to issues of environmental pollution and damage. This awareness is being translated into directives, standards, laws and regulations which suppliers and users need to adhere to.

Such regulations define essentially technological requirements expressed, subject to interpretation and as far as possible, in quantitative terms, for example parts per million of heavy metals in rivers, or of oxides of nitrogen in automobile exhausts.

Quality, again, provides the mechanism for ensuring that:

a. The product design, including all the aspects of its use, is adequate.

b. The manufacturing processes are under sufficient control to meet environmental requirements.

We all subscribe to the view that 'the polluter pays' but to what extent should the supplier's quality management controls extend into the use of his product?

Product Liability

As a consequence of increased consumer expectation for safe products the EU now has strict laws on product liability. The onus is on the supplier to demonstrate that his product was designed and constructed to a sufficient level of quality so that if a user suffered injury it must have been because of misuse rather than because of a poor quality product. Since the law permits a claim to be made up to 10 years after the supply of the product, it can be difficult for a supplier to mount any defence against a product liability claim.

The only rational route for the concerned manufacturer is to have a well-controlled and well-documented approach to the management of quality from all stages from design through material selection and purchase, manufacture, inspection or test, delivery, installation and servicing. This should first of all lessen the risk of any injury occurring because the product quality is well controlled. Secondly, if an injury to a user does occur, documentary evidence is available to demonstrate that the product was correctly designed and manufactured. A good quality management system and good quality records are essential for a supplier of products in the EU. The premium for product liability insurance is likely to reflect the level of quality system in place as well as the more obvious factors relating to perceived risk.

The quality emphasis here is on good documentation and records.

In general, therefore, we can see that Quality Management can help to provide solutions to the four major business issues identified, and can do so by operating at a number of interrelated levels:

a. By providing an overall management philosophy and managerial controls to minimise waste and thereby continuously improve efficiency.

b. By providing managerial and technological controls to ensure that products are adequately designed <u>and</u> supplied to the design requirement with minimal variability.

c. By providing documented controls and records to clearly demonstrate achievement of the required product quality.

Definition of Terms and Related Concepts

Definitions

The sensible study of any subject needs an agreed terminology. ISO 8402 gives a set of definitions which are used as a matter of course in the various national and international standards:

QUALITY

The totality of features and characteristics of an entity that bears on its ability to satisfy stated or implied needs.

QUALITY ASSURANCE

All the planned and systematic actions implemented within the quality system and demonstrated as needed to provide adequate confidence that an entity will fulfil requirements for quality.

QUALITY CONTROL

The operational techniques and activities that are used to fulfil requirements for quality.

QUALITY POLICY

The overall quality intentions and direction of an organisation with regard to quality, as formally expressed by top management.

QUALITY MANAGEMENT

All activities of the overall management function that determine the quality policy, objectives and responsibilities and implement them by means such as quality planning, quality control, quality assurance and quality improvement within the quality system.

QUALITY SYSTEM

The organisational structure, responsibilities, procedures, processes and resources needed to implement quality management.

The difference between quality assurance and quality control should be noted as people are often unsure of whether there is any difference between them.

Quality assurance can be thought of as the assurance that the customer, or a regulatory authority, receives that the quality of a product or service complies with requirements. Quality assurance, therefore, requires evidence that the supplier has complied with requirements. This is why quality assurance has always been very oriented towards the concept of documented systems giving clear evidence to provide the required assurance.

Quality control, on the other hand, means the actions taken to control quality and hence provide assurance. Quality control covers activities such as inspection, testing and process control measurements plus management issues such as feedback and corrective action.

The interrelationship between these concepts can be illustrated as follows:

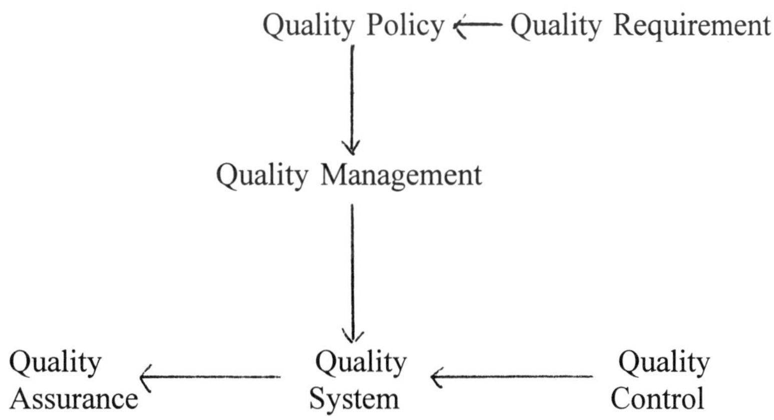

A number of points are worthy of note:

a. Quality Policy is set by top management.

b. The Quality System is the visible expression of the quality management activity.

c. There is no explicit definition of Total Quality Management (TQM).

The emphasis of the approach is, therefore, to stress the importance of the management system and since the ISO standard for a management system (the EN/ISO 9000 series) requires a system to be documented, the philosophy which emerges is that of a well defined and documented approach to management which should give consistency of performance.

The potential danger of this philosophy is that it can restrict creativity and the incremental improvements in processes and procedures which are the route to continuous improvement.

The definition of total quality management (TQM) is not as generally agreed as the other terms. We could explain this by saying TQM is the all embracing term which includes all the other quality related definitions.

One definition which has been produced is:

TOTAL QUALITY MANAGEMENT

Management philosophy and company practices that aim to harness the human and material resources of an organisation in the most effective way to achieve the objectives of the organisation.

This typical definition treats TQM as a business philosophy rather than a set of rules and this is perhaps the best way to consider it, although it is usually quantified within an organisation.

More explanatory definitions of quality

The official definitions for 'quality' are not always easily interpreted for particular circumstances. In particular, they are not readily seen to address the different aspects of quality in a business sense which have been introduced.

Quality has been officially defined as "the totality of features". This is correct but not always helpful. In any particular circumstance, it is true to say that the definition of 'quality' is unique.

But, to determine that unique definition and to help shape an organisation's management approach to 'quality', we need to give some guidance on how "the totality of features" is interpreted and what are the key features in a particular case.

> We have already defined quality as : "The elimination of waste"

In the supply of any product or service we may not be very efficient and waste time, materials or energy, as a result of poor quality materials or work (need for repair or rectification) or poor organisation (waste of time and energy in the business processes). Elimination of waste may not improve the inherent key features of a product but may well increase the quality of service (better delivery, lower cost, more consistent performance) and thus the value to the customer. This definition of quality is particularly relevant when supplying to a defined specification.

> Another definition of quality is : "Delighting the customer"

This means providing a product or service which exceeds the customer expectations, with better performance, better service, faster delivery, better value for money. This definition applies particularly to the supply of consumer products in a competitive market where there is no defined customer specification, but only market predictions, e.g. motor vehicles, cameras, hotel service or air transport service.

We can see, therefore, that the definition of 'quality' needs a different focus for different circumstances and that the ISO 8402 definition, whilst correct, is not sufficient.

Another way of helping to define quality in a particular set of circumstances is to recognise that quality has four features:

1. Conformance to specification.
2. Conformance to legal and statutory requirements.
3. Meeting the anticipated wishes of the customer.
4. Beating the competition.

The relative importance of these different features will vary for different products or services but, in all cases, focusing on the requirements of these four features can provide direction on the specific definition of quality. Usually <u>all</u> of these features must be satisfied.

This discussion of the different meanings of the word 'quality' supports the view that TQM is a broad concept and that the ISO definition of terms introduced earlier, whilst correct, can easily be interpreted in too restrictive a manner.

Related concepts

There are a number of management techniques which are frequently used to support concepts of quality management and are sometimes included under the broad banner of TQM.

Just-in-time (JIT)

Holding of inventory is expensive because it ties up warehousing space and finance. It also reduces flexibility, since the manufacturer must assemble product using existing inventory and/or will 'manufacture for stock' to minimise production costs.

The idea of JIT is that a supplier holds minimum (almost zero) inventory but plans for deliveries of raw material or components just when they are needed. This gives him flexibility to satisfy changing or different customer needs and reduces the cost of holding inventory. It requires a new approach to production planning and very good relationships with suppliers - their production planning must also be based on their customer's plans. In practice, they should also be geographically close.

This is a concept which fits the 'reduction of waste' idea and the need for competitiveness.

The concept, although of universal applicability, has mostly been implemented by assemblers of consumer products, such as automobiles, consumer electronics, where competitive price pressures are high and there is need to have a flexible response to the market.

Concurrent engineering (CE)

This concept, sometimes known as 'simultaneous engineering' has been devised, like JIT, to meet the challenges of competitive pressures and rapidly changing market needs.

CE is a concept to reduce new product development time by simultaneously developing engineering and production methods during the design stage. This can lead to faster new product development and requires the use of multi-disciplinary project teams to develop new products, together with very good communication links within the supplier organisation.

The role of IT-based management information systems can be valuable in this approach.

Quality function development (QFD)

This is a concept for logically analysing quality needs and for correctly assigning priority to quality tasks.

It requires a team approach, takes customer needs as the starting point and translates them into technical requirements of design and manufacturing via a series of logical steps which can then be analyzed in terms of the priorities for a quality programme.

Essentially, QFD is a disciplined, formal analysis of customer needs.

Failure mode and effect analysis (FMEA)

This analytical technique is used in design and is particularly popular in the automotive industry. It identifies the ways in which a product could fail (the mode) and examines the likely consequences of failure (the effect). Different weightings are given to failure severity, and the aim is to rank the different failure modes in order of criticality so that the most critical can be eliminated first.

Benchmarking

Companies can use benchmarking techniques to improve the quality of their performance by comparing each business process in a company with best practice worldwide and seeking to emulate it. Comparisons are by process e.g. order taking, delivery, manufacturing throughput, rather than by company. This enables benchmarking to be undertaken with companies in different industry sectors, if accurate benchmarking of a competitor is difficult because of access to data.

There are a number of other, specific, tools which will be discussed in a later section. It is noteworthy that most of these concepts have been developed as a way of coping with competitive pressures in the supply of manufactured products (principally consumer items) and it is in these circumstances that they are primarily used.

2.2 Quality Management of Industrial Processes

The basic aim of modern quality management is to get things right first time. It is wasteful in time, money and materials to repair, rectify or scrap and re-make, therefore the old-fashioned concept of design, manufacture then inspection is not good enough. The modern concept is correct design and controlled manufacture so that correct product is supplied.

Inspection cannot be entirely discarded but it should not be seen as the controlling activity in a quality sense.

Product, Process and System

As we have seen, the total quality approach to management is directed towards consistently supplying the ultimate customer with what he wants, when he wants it, at an acceptable price to him and at an economical cost to the supplier.

The delivery, price and performance of the product are of prime importance to the end customer, but most customers, except those who actively foster a close (collaborative) relationship with their suppliers, are probably not too concerned about *how* these features are achieved. However, this is of vital concern to the supplier since the economics of this aspect will largely determine the extent of his future business success.

The supplier must therefore be aware of the process, or sequence of processes, which produces the end-product. For this to happen effectively, there must be a climate of good management practice; effectively this is TQM.

Existing Situation

The organisational sequence of events that usually applies is shown as follows:

Management philosophy
↓
System
↓
Process(es)
↓
Product(s)

Senior management will have set company policy regarding quality. Within this, a system will exist to run the sequence of processes needed to produce the product. Such a system will contain at least the principles set out in standards such as ISO/EN 9000.

<u>THE PRODUCT</u>

This is what the customer is buying.

It must have the features he wants (at least).

<u>THE PROCESS</u>

The means by which the product/service is produced/delivered.

For economic provision of what the customer wants, there must be knowledge of relationships between process inputs and outputs.

<u>THE SYSTEM</u>

The management mechanism which sustains the process(es) to provide the product required by the customer.

The three elements - product, process and management system - though distinct, are all closely interwoven. This can be expanded as follows:

The Product

What product or service does the customer want, need or expect? It is a management responsibility to know this so that profitable manufacture or service can be arranged. The customer's requirements must be linked to the capability to provide for them. This is the essence of the approach known as quality function deployment (QFD).

Given the product requirements (driven by the customer's wants), it is important to ensure that potential problems in performance of the product and its ownership, have been considered and addressed. This approach is the basis of 'failure mode and effects analysis' (FMEA), which in practice is similar to a safety investigation to identify hazards and assess risks.

There should be monitoring of the customer's experience with the product so that any problem areas can be identified, controlled and if necessary improved.

The Process

Data about product characteristics and performance, both from field experience and from in-house testing, give valuable information about the process. The overall process includes design and manufacture and all the associated activities. Here the purpose is to try to establish cause and effect relationships between the way the process is operated and the properties of the product it produces. As process knowledge develops, so the potential grows to run the process in order to guarantee the required product features at minimum cost.

Processes should be monitored, both in their inputs and outputs so that they can be run in a stable, controlled manner. A further desirable step often is to improve the process operation beyond its present capability. Process capability studies are beyond the scope of this book.

The System

This is what holds all the quality activities together. A quality culture will ensure that all staff have the appropriate skills and attitudes to work together as necessary to run the processes to produce the product that will delight the final customer. It also implies and in fact requires that all staff will co-operate with each other.

The operation of the system should be monitored so that it can be controlled to remain effective. Not only that but any scope for improvement should be actively sought.

Monitoring, Control and Improvement

It is implicit in a total quality approach that there should be never-ending improvement. However, improvement can only be implemented effectively in a controlled environment. In turn, control has to be worked for and it is usually necessary to monitor events, first of all to detect lack of control, then to achieve control and finally, to maintain control.

As we have just seen, these actions of monitoring, controlling and improving are applicable to the product, the process and the system.

Monitoring

This is the observation on a continuing basis of what is happening. Monitoring adds cost but no value and may save cost at a later stage. Inspection can also provide the intangible feel-good factor to suppliers and customers and is the basis of much quality assurance.

The essence of monitoring is to look at trends and changes (or lack of them) *over time* to reveal actions necessary to be taken with processes and the system, or to confirm that all is well.

Control

Monitoring on its own is not enough. When it provides evidence that the situation is not acceptable, management action must be taken to make it so. This does not necessarily mean that it has to be senior management who are involved. The total quality principle is that it should be whoever is available with the requisite skills and authority closest to the decision point. The feedback loop, and hence the time to correct things, should be as short as possible.

Control implies that the situation being monitored is known to be capable of operating in a stable, predictable mode and then actions are taken as necessary over time to keep it in this mode.

Improvement

Even though processes may be maintained in a stable condition and therefore be in control, the output may not be good enough. For example, the characteristics of a product may all lie within specification though some items may be close to the upper or lower values specified. Although the process can be run at will in this state of control there is usually a need to tighten specification limits or improve the specified values. Commercial needs (competition or customer expectations) or efficiency needs (reducing waste) are likely to arise. In order to meet these, the product items would have to be made with smaller variability amongst themselves as a group. It is then a matter of process improvement to obtain this required greater uniformity of product performance.

In general, improvement is harder to achieve than is the initial state of control but when it has been, the new level of performance must be subject to monitoring and control. And so the cycle repeats, aiming for better quality at minimum or reduced cost.

2.3 Tools and Techniques Used for Quality Management

Tools and Techniques

We have seen that management must play a crucial role in implementing a total quality culture. High profile leadership by example, underpinned by a system based on sound principles, is a key requirement for success in delivering the quality of product and service appreciated by the customer.

Assuming that management provides the leadership necessary, what should actually be done?

> People must collect facts about the operations with which they are associated, take improvement actions on the basis of these facts and ensure that all new information is communicated to everyone who can make use of it.

Some facts are blindingly obvious but most improvement opportunities, especially the later ones, have to be investigated in a deliberate, systematic way. For this, various tools and techniques, mostly statistical, have been developed. According to Chambers' 20th Century Dictionary, a tool is a 'working instruction, an accomplice or anything necessary in the pursuit of a particular activity'. Technique refers to a 'method of performance or manipulation especially anything concerned with the mechanical part of an artistic performance'. So, officially, a tool is what you use and technique is how you use it. In practice both terms seem to be used to mean tool.

There are two types of TQM tool. First there are the management devices used to get TQM started, growing and maintained. Then there are those tools used to collect the facts needed to make things happen in terms of quality improvement. The first group are largely motivational techniques.

The second group of tools is predominantly, though not entirely, statistical in nature. This seems to produce an element of fear in many people. So much so that often the available tools will either not be used at all or will be used in such a mechanistic way that the data are not made to yield all the information that they contain. At best, this is inefficient but at worst, worthwhile improvement opportunities may be missed or delayed in implementation. This is a pity because statistical methods are only the tools needed for interpreting numbers and in particular, numbers obtained from samples.

Simple tools can make a vast improvement to an organisation's knowledge about its processes whether these be in production, administration, service or other areas.

The process knowledge which is most useful and can be profitably exploited is that of relationships between inputs and outputs. Once these relationships are known, the best output properties obtainable with the existing process can be ensured by paying careful attention to control of the process inputs.

Useful Tools

It is convenient to group tools into three sets: basic, intermediate and advanced.

Basic:	7 basic tools.
Intermediate:	Sampling, experimental design.
Advanced:	Operational research.

The simple basic tools are the ones that have been championed by Ishikawa in particular and referred to as the seven basic tools:

Flow chart
Pareto diagrams
Check sheets
Cause and effect diagrams
Histograms
Control charts
Scatter diagrams

Ishikawa maintained that people are a company's key resource and should be treated with respect. He also stated that *all* company personnel, senior management included, should be trained to know how to use the above tools. This training, which should include an element of education, is most effective when carried out in real situations familiar to the participants.

Although simple, the regular use of these tools brings a structure and discipline to problem solving. Being graphical, their message in any given case should be readily understood by anyone else in the organisation. It seems to be generally true that a company can make a big leap forward in its quality programme using only these basic tools coupled with enquiring minds and co-operating personnel.

Enquiring minds are very valuable. They will not only observe that problems exist but will try to identify the causes and propose a solution in very much the same way as a doctor, when faced with a sick person, will work from the symptoms, the patient's answers to questions and his own experience towards a likely cause of the trouble and hence prescribe a remedy. This is an example of process control (attempted, at least). First the person is observed to depart from normal health, then the features of this departure are used to assess the reason for it and then a corrective action is applied with the aim of restoring the normal state of health. The patient will generally continue to be monitored to check on the effectiveness of the corrective action. If inadequate, further corrections will be applied. Even when restored to normal health, the patient will (subconsciously) continue to monitor his health and when it deteriorates by more than an acceptable amount, an out-of-control signal will be generated to consult the doctor again (that is, to identify the problem and eliminate it, so returning to the controlled state).

In a similar way, a process can be monitored through measurements made of process variables (temperature, speed, vibration, raw material properties etc.) and of output product properties. Ideally, these measurements should be made using the principles of metrology and with calibrated equipment. Sometimes this is not possible and rough and ready measures have to be used. These are likely to be better than nothing.

When an observed measurement, or a group of them, deviates from expected by more than some previously specified amount, it is a signal that the process output is not what it was before, probably because one or more of the process input conditions have changed. The available evidence (nature and size of the output deviation, process log, talking to people, experience) can all be used to identify the likely reason, make the appropriate adjustments and bring back the product characteristics to their required values. This is process control.

Other 'Scientific' Tools

The intermediate group of methods includes sampling methods, parameter estimation, hypothesis testing and basic experimental design. These concepts are a little more involved than the basic tools just described but properly introduced, they should hold no terrors. In essence, they introduce some more rigour and objectivity into process investigation and the drawing of conclusions.

Advanced experimental design, multivariate analysis and the tools of operational research form the most sophisticated set of tools. They are concerned usually with matters of optimisation and, in most cases, will be used only infrequently. Most organisations, except those with their own in-house expertise in this area, will normally need to call on outside advice.

The important point is that the correct tool should be used in a given case but in the early stages of running projects in a developing TQM environment, simple projects needing only simple tools should be selected for study. Early successes are the best guarantee of future progress.

The Techniques

In addition to these statistical tools, there are various techniques used to enhance internal quality consciousness and quality performance. These are analytical or problem solving tools and can be listed as:

Benchmarking
Nominal group technique (a variation on brainstorming)
Time charting and analysis
Six sigma and sixth sigma
Five whys
Cost of quality
Quality Function Deployment (QFD)
Failure Mode and Effect Analysis (FMEA)
Force field analysis
Value engineering
Soft systems methodology
Moments of truth.

Some of these are so simple as to seem almost trivial and hardly worth including in such a list. However, anything which can help to get an organisation moving on the road to total quality is to be made full use of. They have therefore been included on the basis that the simple approach is often the most effective.

Benchmarking is intended to be carried out on a systematic basis by comparing, as rigorously and ruthlessly as possible, one's own performance against that of the best of the competitors and also against that of the best in any field. There is always something that can be learnt to lead to improved performance.

The nominal group technique is a variation on traditional brainstorming, there being an initial period for private contemplation of ideas which are then made public in the usual round-robin way. The ideas are then ranked by each team member in private and a consensus view is obtained without any one team member having been able to dominate proceedings (provided that the facilitator has remained in control to apply the rules of procedure).

Time charting uses flow charts to map activities showing where time is spent (and wasted) and where alternative methods might reduce it. The idea is to highlight where improvement opportunities exist.

'Six sigma' focuses on extreme departures from a norm for some quality attribute. In the first place this requires that some performance measure, however crude, can be constructed for the activity. The people involved monitor their own performance through a time plot of the measure, taking note of any undesirable high or low values. In this way, a 'feel' for likely causes of adverse performance can be built up. If valid control limits based on statistical theory can be constructed, so much the better, if not, the plot can indicate step changes, trends and cycles in the activity provided that management assists in interpreting the plot and does not cast blame for what might be only part of the 'noise' in the performance measure.

'Sixth sigma' expresses the idea that if sometimes quality performance is very extreme (that is good, say) then why can it not always be like that?

The 'five whys' merely observe that if the reason for some activity is questioned, then a supplementary question is usually generated and after about five such iterations, an improvement opportunity will often be obvious or perhaps a further promising line of enquiry will be indicated. Such a simple approach can lead to the cause of a difficulty being traced rather than trying to make progress by treating the symptoms.

The cost of quality is a big topic and very much has been written about it. It is an attractive proposition to evaluate the costs associated with defect prevention, with quality appraisal and with quality failure costs so that these can be compared to the 'benefits' of a quality programme. After all, this expresses quality in the language that all managers are familiar with, that of money. However, some external failure costs are, in Deming's words, "unknown and unknowable". What is important is to be able to monitor trends in costs related to quality issues and to know their location. Organisations following the total quality route report that dramatic savings are not usually observable against specific improvement projects, but rather there is a gradual increase in profitability over a timescale, often of several years.

Quality Function Deployment (QFD) is a technique used to relate customer needs to the technical features of a product or service. The essence is to rank the customer needs and identify which are the most important technical characteristics that must be considered in catering for these needs. The assembly of a multi-disciplinary team to carry out such a design activity is frequently found to be one of its main strengths; it focuses on the inter-relationship of all departments involved and by doing so, can lead to shorter product development times and a product which is easier to manufacture, works successfully the first time and satisfies the customer. All of these desirable features are likely to lead to overall reduced cost.

Failure Mode and Effect Analysis (FMEA), and variants of it, is a structured approach to identifying the ways in which a product or service could fail (the mode) and examining the likely consequences of such failures (the effect). Weightings, of necessity usually subjective, are given to the severity of any given failure, the chance of it occurring and the chance of it not being found before release into the market place. These scores are combined into one value, either by adding or multiplying them, and then they are ranked in order of size. The highest scores suggest which possible failure modes should be eliminated first.

Force Field Analysis has something of the cause and effect diagram about it, coupled with the use of brainstorming. Basically, each force pointing towards change is attached to a right arrow, say, and each force which is against change is drawn alongside a left arrow. The dominant forces are identified and those which can be changed are highlighted and a programme for action to change them is formulated. Yet again, it is a technique which fosters the team approach from which probably comes much of the benefit.

Value Engineering has traditionally been used in engineering design as a cost reduction technique. Its main thrust is to identify the important features of a product or service and then to use so-called 'creative thinking' in identifying alternative ways in which the function that the product or service provides may be delivered. The group approach used can also facilitate implementation of the chosen alternative.

Soft Systems Methodology is applied to situations where there is little or no discernible structure and an absence of suitable data. It can be used to tackle 'woolly' problem areas and essentially works by concentrating attention on the inter-relating factors that affect the system. An improved understanding, however sketchy, of what to do next about the system can emerge.

Moments of Truth refer to those times when a customer is in contact with any aspect of the organisation with which he/she is dealing. It is at such times that there is the potential for the customer to become dissatisfied and, at worst, then take his/her business away. Perhaps that is not the worst - they might also sue for damages, and win! The trick here is to put oneself in the place of the customer and examine all likely contact points with one's organisation and assess the possibilities for something fouling up. An early example of this is the suggestion to senior managers that they should telephone their organisation from outside to see how their switchboard answers. This is important since the operator may be one of the lowest paid staff but is in a key position to influence customers. In an extreme case, a real potential customer might be lost forever without anyone ever knowing - part of the 'unknown and unknowable' cost of quality (or lack of it).

3 QUALITY MANAGEMENT, APPLIED TO ADHESIVE TECHNOLOGY

3.1 The Special Characteristics of Adhesive Bonding

In many ways, adhesive bonding is just another manufacturing technology such as welding, bolting, forming, machining or casting. Therefore it is subject to the same rules for process capability analysis, process monitoring and, hence, process control. There is no reason why the standard quality tools and techniques and, in particular, the statistically based tools cannot be used.

However, adhesive bonding has some special characteristics which make the quality management and hence, quality assurance, of the process more difficult. The first characteristic is that adhesive bonding is used in a very wide range of applications, from packaging to aerospace, with very different needs and a wide range of adhesive types and methods of bonding. Secondly, the technology itself is complex with many process variables and the interaction between them not fully understood. Thirdly, inspection methods, particularly non-destructive methods, are not often very indicative of joint performance. For these reasons the application of the standard techniques requires more data than is often available, so a slightly different approach to quality management and quality assurance is needed.

To provide factual information and a basis for development of a model, real industrial examples of the use of adhesives in manufacturing industry have been documented, with the cooperation of participating companies, plus others. Information consisted of the following:-

a) product development

b) process development

   i.e. a company's design process

c) an active manufacturing process, resulting from a) and b)

Analysis of manufacturing processes

Three assembly processes from Swedish industry are compared in Figures 3.1.1 - 3.1.3, courtesy of IVF, Sweden. They were selected because their methods of adhesive application had different degrees of automation and the requirements of the quality of the joints were different. Key points are summarised below each figure.

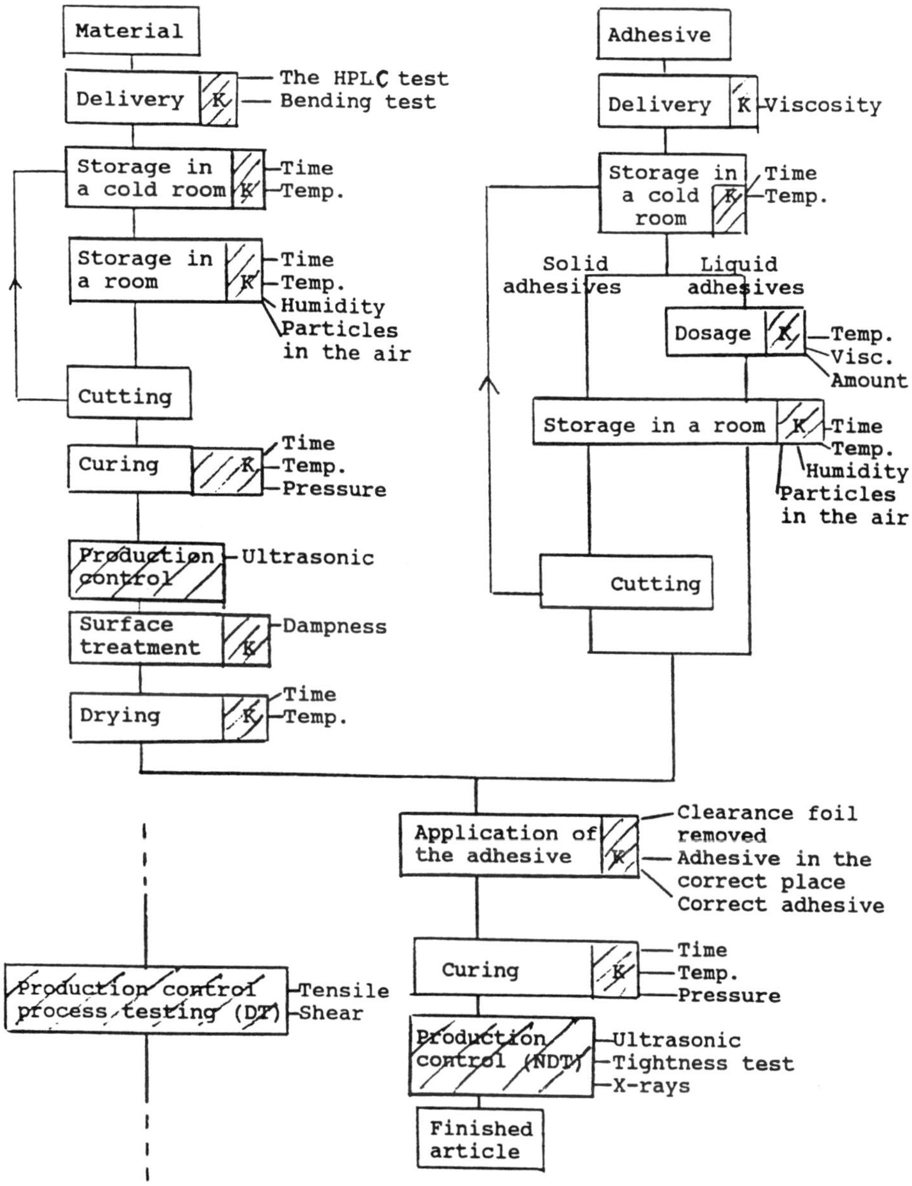

Saab-Scania	- carbon-epoxy composite / film adhesive
Key considerations	- material surfaces, adhesive quality
Process controls	- control of air humidity and particulate content
Final controls	- extremely important
	- ultrasound, x-ray and tightness tests
	- destructive test of parallel processed test pieces

Fig. 3.1.1 Aircraft Wing Spoiler [MD80]

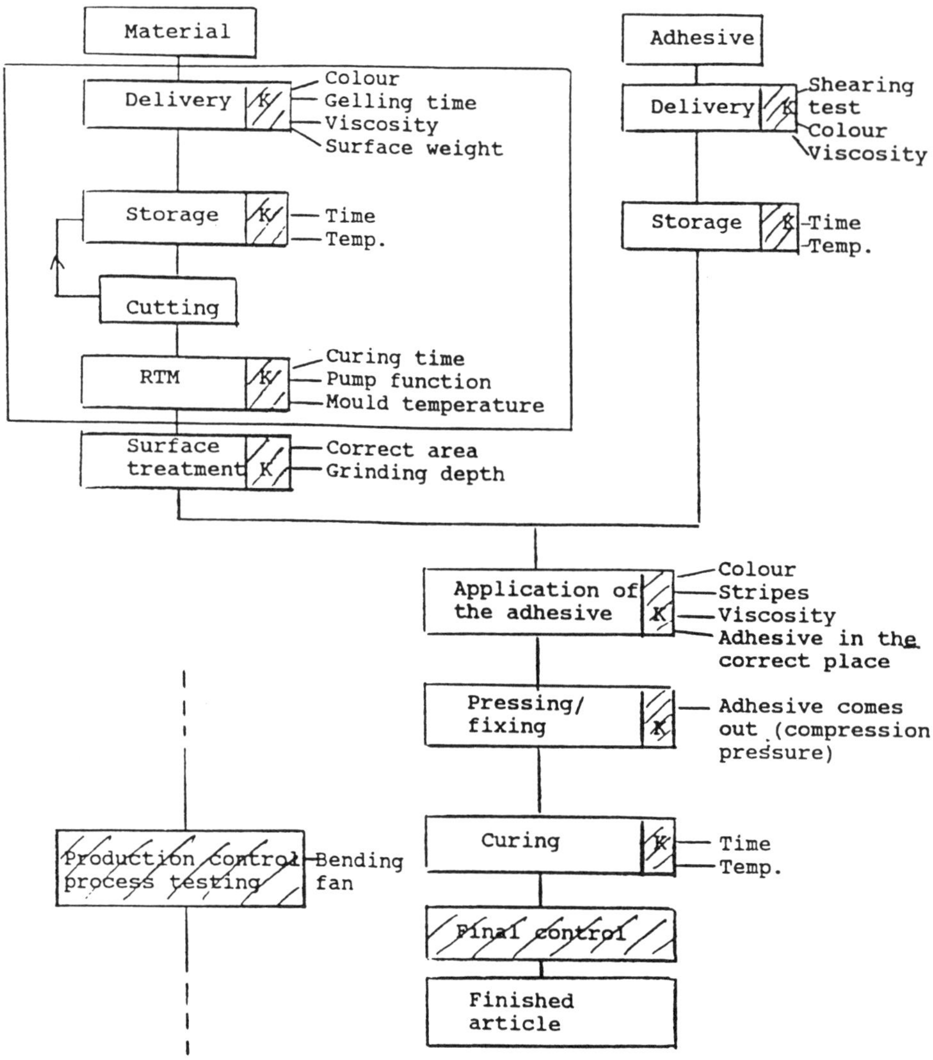

Borealis Industries	- glass fibre-polyester composite / two-part adhesive
Key considerations	- process controls on panels are not for bonding, but for surface finish or strength
	- joints oversized to allow for incomplete filling with adhesive
Process controls	- adequate mixing (no stripes, colour)
	- correct placement of adhesive
Final controls	- tightness important, therefore checks made that adhesive runs out at pressing/fixing stage
	- production control process testing

Fig. 3.1.2 Off-road Vehicle Panel

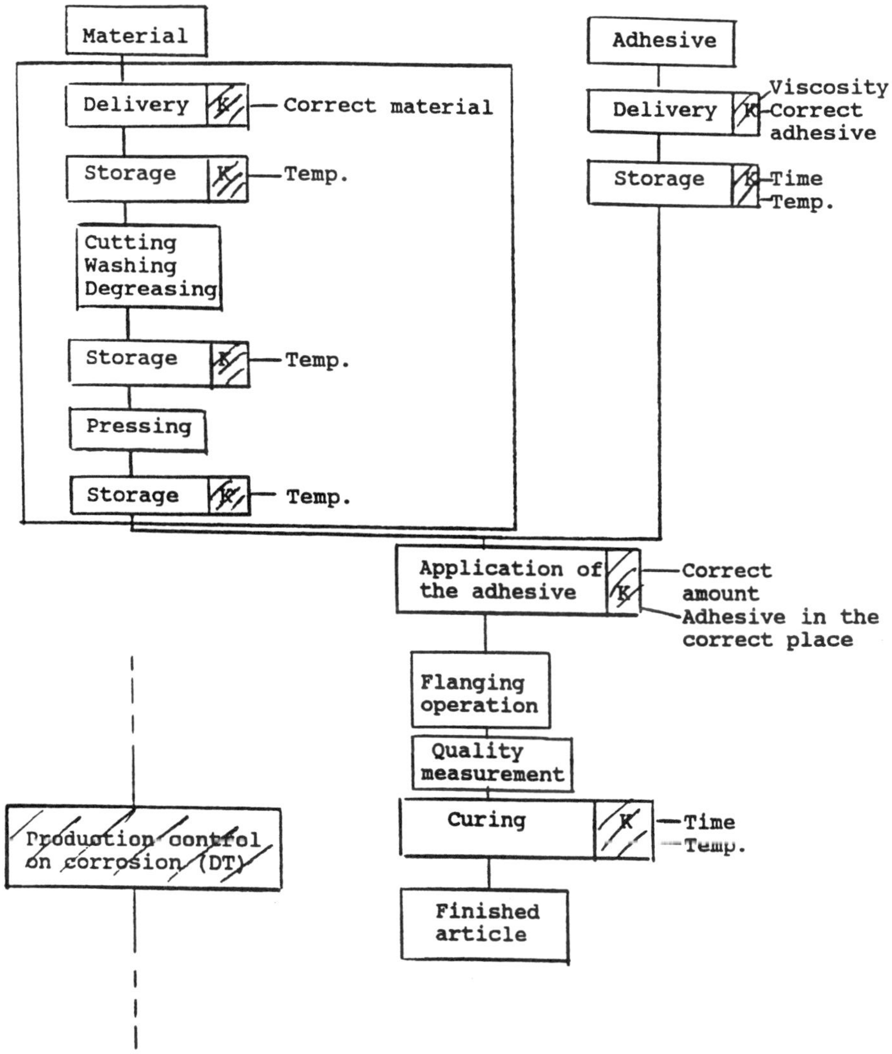

Saab-Scania	- steel / paste adhesive
Key considerations	- applied during flanging operation
	- main purpose to act as anti-corrosion sealant
	- adhesive cured with paint, and cure conditions determined by paint system
Process controls	- optical control of robotic adhesive deposition
Final controls	- joint hidden after flanging and no inspection possible

Fig. 3.1.3 Auto Bonnet, or Hood [Saab 9000]

UK activity started with the analysis of best practice and problems in the quality management of industrial adhesive bonding. This was accomplished by the collection of information from all the participants and by the detailed analysis of manufacturing processes submitted by the industrial companies supporting the project, both end–users and the customers of adhesive suppliers, plus the example of GRP pipe bonding from Delft and other contacts. Manufacturing processes were documented using a checklist shown below in Figure 3.1.4.

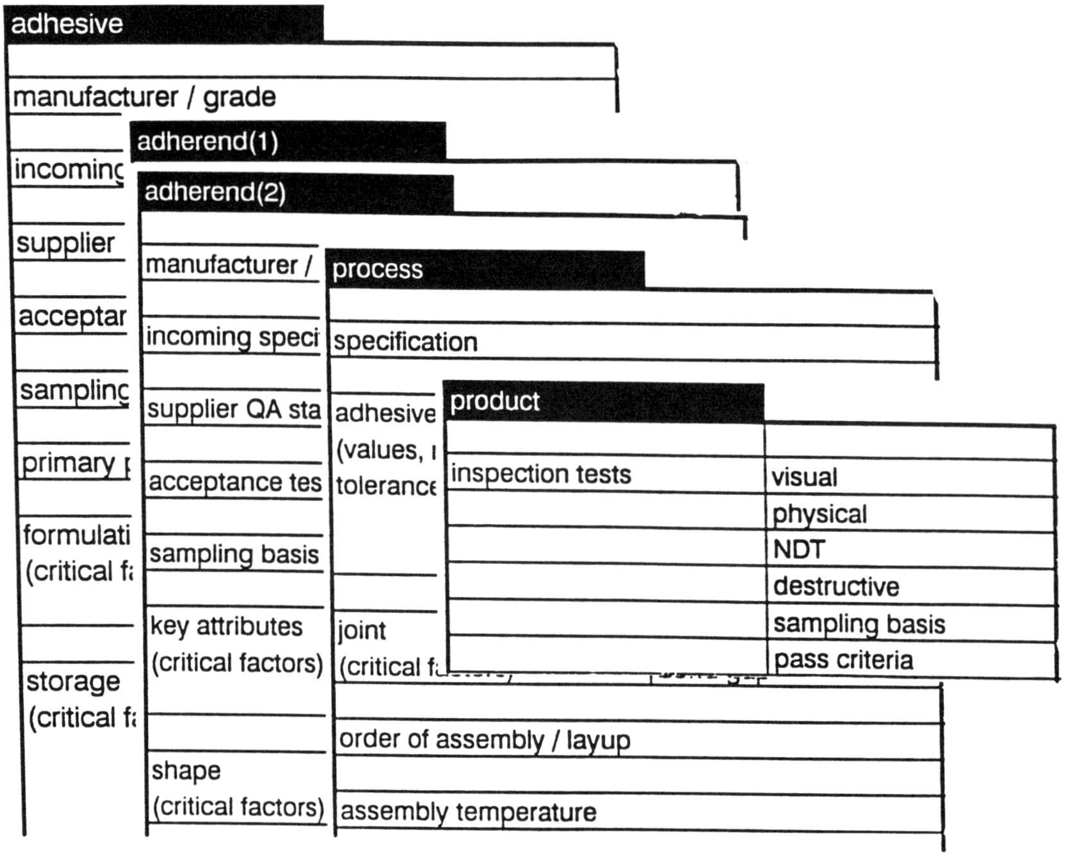

Fig. 3.1.4 Manufacturing process Checklist

Many of the documented processes are collected in APPENDIX A. They cover a wide range of adhesively bonded products from many industrial sectors. The purpose of this effort was to obtain all the critical factors and control points about a particular bonding process, to direct the development of a generic quality management model with the optimum control measurements.

A blank copy of the checklist is included for future use, both at the end of APPENDIX A and on the QUASIAT disks (filename 'PROCESS.WB1'), accessible to users of Borland Quattro Pro 5.0 for Windows, or equivalent spreadsheet software.

An analysis of the documented manufacturing processes showed that some activities are monitored in great detail and others only a little. Sourcing and storage of materials or adhesives were highly detailed, because much information was available from suppliers. On the other hand many aspects of assembly, which were the internal responsibility of the end-user, relied on in-house experience and were more of an area where guidance is needed. Final inspection had, not surprisingly, a low frequency in the analysis because it was the inherent difficulty of this activity for bonded products which provided much of the motivation for this programme.

A broad cross-section of adhesive usage is summarised below in Table 3.1.1, which shows the variety, the differentiating points and the common features. It was an early attempt to define the critical design, assembly or performance requirements for any adhesive bonding process. It helped to define the key differentiating factors in a manufacturing process (production rate and level of mechanisation). High/low variants are included in the table and their place in the overall model is described in the next section. Although the technical aspects of control are very different in each application, a standard plan for quality management can be used, but it must be differentiated by factors other than merely industry sector.

Table 3.1.1 Key Points of a Range of Adhesive Bonding Applications

INDUSTRY	AEROSPACE	AUTOMOTIVE	PACKAGING	CONSTRUCTION
PRODUCT	wing vane	headlamp unit	flexible lamination	joint sealing
ADHERENDS	composite/aluminium	glass/polypropylene	PET/LDPE	concrete/u-PVC
ADHESIVE	epoxy film	butyl hot-melt	2-part polyurethane	2-part polysulphide
JOINT TYPE	structural	semi-structural	non-structural	sealing
THROUGHPUT	4-10 / week	1-2 / min	100-300 metres/min	30 metres/day
LOCATION	factory	factory	factory	site
SKILL BASIS	craft/system	system	system	craft
CONTROL SYSTEM	process control	process control	process control	training and trust
FINAL INSPECTION	visual and ultrasonic A-scan, every item	pressure test, 100%	visual and peel test, 1 reel/shift	visual and qualitative adhesion, hardness
production rate	low	high	high	high
level of mechanisation	low	high	high	low

Analysis of product and process design

The design stages of both product and assembly were also studied with the participating companies, including several specific developments which had led to manufacturing processes already documented. The objective here was to analyse the implications of design on quality management, or vice versa.

A concise summary of this task was not possible, but analyses of both product and process design highlighted differences between end-users, such as when the joining of parts is considered. Sometimes the use of adhesives to join the product was considered much earlier (in product design) than the, equally important, implications for production of the use of adhesives (in process design), as shown in Figure 3.1.5 below. The control points, methods and values could be defined by different groups, such as design, production, quality assurance or the customer. At best, the quality management methods used to impose reliable manufacture were totally integrated with a company's quality system. Real case studies emphasised one of the major points of simultaneous engineering; that effort spent prior to production is necessary, most valuable and cost-effective.

Nissan UK define this procedure as Pre-Production Quality Assurance (PPQA), when the aim is to identify and control all the key factors in development and production and to ensure that all problems are resolved before volume production.

Early PPQA - Product Development	Timing Contingency plans Installation trials Design FMEAs Design testing
Second Phase PPQA - Process Development	Training Gauging Capability studies Foolproofing (Poka-Yoke) Workshop management

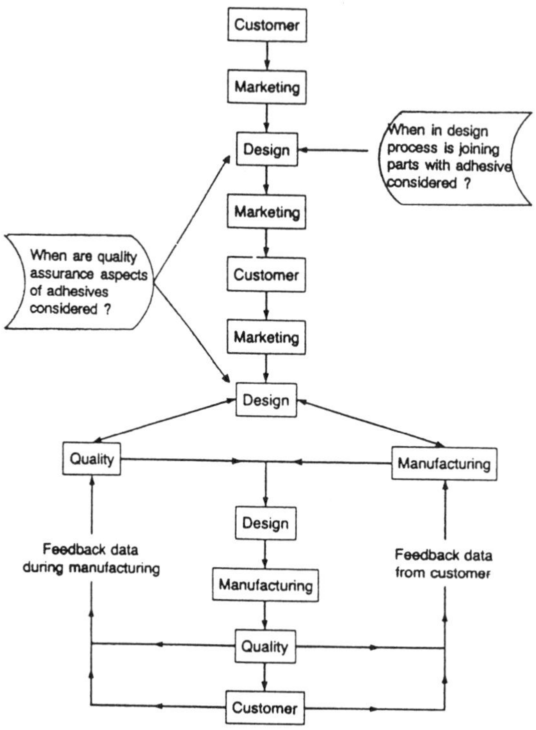

Fig. 3.1.5 Potential divergence between design decisions - to join with adhesive vs implications for quality

3.2 The Quality Management Model, Text-based

Adhesive bonding is an important joining technique which is used in a wide variety of industries (packaging, automotive, aerospace, general structural, building and construction are key ones). In many products the performance and reliability of the adhesive joint is critical to the product's performance and, sometimes, to its safe operation.

The problem of assuring the quality and reliability of an adhesive joint is the absence of suitable non-destructive examination methods, in all but a few special cases. Therefore, inspection of adhesive bonds is not a feasible way of assuring quality and in modern manufacturing philosophy, it is after the event and is an unnecessary activity in a well-controlled process. The only way to provide assurance is to systematically manage and control the whole operation from design of the joint through to final assembly. In this way, the possibility of poor quality joints being produced is reduced to the minimum because proven procedures are being followed at all times.

Every adhesive bonding situation is unique but all adhesive bonding situations have a number of common features. Consequently, this generic model is designed to cover all situations. All the major stages from design through to final assembly and inspection are listed which define the requirements, methods of quality verification and the remedial action at each stage. Each stage can be accessed individually, progressively in more detail leading ultimately to the identification of the specific measures needed for an individual application. The model therefore acts as guidance to the user in defining the specific quality requirements for a joint and the specific quality management actions needed to ensure the quality requirements are met.

General Approach

To accommodate the varied products, adhesive types and adherend types, we need to structure the generic model into 3 layers (from the very general to the very specific):

Layer 1 - Generic Model
Layer 2 - Manufacturing route related sub-model
Layer 3 - Specific application

Layer 1 is completely general and can be applied to all situations.

Layer 3 is entirely specific to a particular user for a particular joint or component.

Layer 2 is, therefore, the key layer as it is the interface between the general approach and the specific user.

Development of Layer 2

In simple terms the sequence of making an adhesive bond is as follows:

Design of joint
|
Selection of materials — Design activity
|
Selection of adhesive
|
Pre-treatment of surfaces
|
Assembly
| — Manufacturing activity
Cure
|
Final inspection

The quality management of the process relies on two major concepts:

1. Control of joint design and specification of materials and processes.
2. Process monitoring and/or inspection.

The application of these concepts relies on the definition of the requirements (unique to each situation), the appropriate information flow/decision making process (shown in Figure 3.2.1) and the use of quality management tools and techniques at the appropriate point in the process (see Figure 3.2.2).

The key determinant of the quality of an adhesive joint is the design of the joint. Design in this context includes material selection (both adhesive and adherends) and the method of joining - in other words the complete specification of the joint. Unless the joint design is completely specified then there is no basis on which to use process monitoring and there would have to be reliance on inspection of the completed joint, which is known to be unsatisfactory. Therefore the quality management of the design is of primary importance

The process monitoring and control of the bonding process is clearly dependent to a very large extent on the type of application. In reality, it is the manufacturing activity which will require differences in the model. The key features of the manufacturing activity which determine the quality management model are:

1. Production rate.
2. Level of mechanisation of the process steps.

These, in their turn, are affected by product type, adhesive and adherend type and component design. The criticality of the joint is not a factor. Clearly, the more critical the joint the more rigorously the quality must be controlled and assured but that is determined by the detailed application of the model. (Joint criticality is also a factor in overall joint/component design).

From the concept point of view, the assembly process can be characterised in one of four ways and these lead to the four sub-models.

* 1. **High** production rate, **high** level of mechanisation.
e.g. automotive components or packaging applications

2. **High** production rate, **low** level of mechanisation.
e.g. some consumer goods or construction sealing

3. **Low** production rate, **high** level of mechanisation.
e.g. critical space or optical assemblies

4. **Low** production rate, **low** level of mechanisation.
e.g. airframe assembly

* **High** production rate is considered as >10 joints/hour.
Low production rate is considered as <10 joints/hour.

Sub-models 2 and 4 relate to situations where operator skill is a determining factor (i.e. a craft based approach) whereas Sub-models 1 and 3 relate to more systems-controlled situations. It is, of course, possible that different levels of mechanisation may be applied at different stages in manufacture.

The quality-related features of the different levels of mechanisation can be summarised as follows:

High level of mechanisation:	Pre-qualified procedures. On-line measurement and process control.
Low level of mechanisation:	Training. Inspection. Pre-qualified procedures.

The generic model (Layer 1) and the four sub-models (Layer 2) are attached in the form of activity, quality requirement, method of validation/control and corrective action, see Tables 3.2.1 - 5. They should be studied in conjunction with Figures 3.2.1 and 3.2.2. The final, useable model will be an amalgam of these flow charts and figures. How they will be put together is dependent on how the user will wish to 'see' the model.

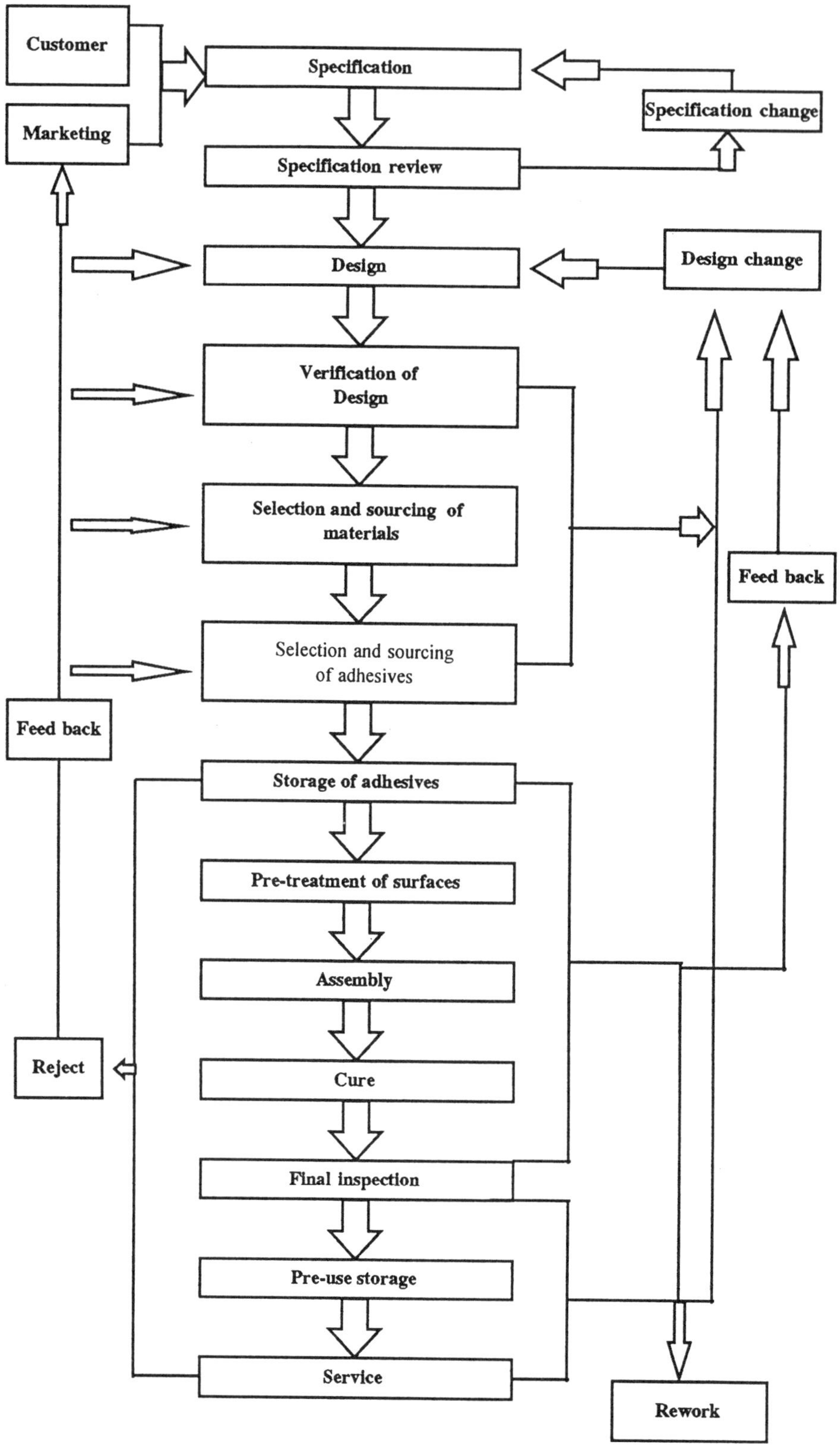

Fig. 3.2.1 Process Flow Chart

Process Steps	Tools and Techniques
Specification review	
Design	Taguchi FMEA QFD
Design verification	Product testing
Selection and sourcing of materials	Product testing
Pre-treating	SPC Validated procedure
Assembly	SPC Validated procedure Inspection
Cure	SPC Validated procedure
Final inspection	Product testing
Pre-use storage	Product testing
Service	FMEA

Fig. 3.2.2 Tools and Techniques, and Their Application

Table 3.2.1 Generic Flowchart of the design and Manufacturing Process, QUASIAT, Level 1

Activity	Quality Requirement	Method of Validation/Control	Corrective Action
1. Specification Review	a. Specification of operating conditions. b. Performance requirements. c. Test requirements. d. Safety requirements. e. Environmental/ statutory requirements.	Review of: Customer spec. Test work Records Published data National and International standards	Specification change.
2. Design.	Specification requirements achieved in an efficient, economic and consistent manner.	Quality Function Deployment. Failure Mode and Effect Analysis. Taguchi loss function analysis.	Design changes.
3. Verification of Design.	Does design satisfy requirements? Is design economic and efficient? Are proposed production methods and test methods satisfactory?	Qualification tests. Alternative calculations. Comparison with conventional fixing methods. Design review.	Design change.
4. Selection and sourcing of materials.	a. Joint design requirements: Strength Weight Environment Appearance Dimensional control. b. Specified component requirements: Strength Weight Environment Ease of forming Appearance Durability Other properties specific to product.	Test records. Supplier certification. Published data. Experience of previous use.	Return to supplier. Re-select. Design change.

Activity	Quality Requirement	Method of Validation/Control	Corrective Action
5. Selection and sourcing of adhesive.	a. Joint design requirements: Strength Compatibility Environment Durability Gap filling properties. b. Production requirements: Ease of dispensing Shelf life Tolerance to environment Cure time/ temperature.	Test records. Supplier certification. Published data. Experience of previous use.	Return to supplier. Re-select. Design change.
6. Storage of adhesives.	Requirements specified by adhesive supplier: Shelf life Packaging Temperature Humidity.	Inspection of packages. Batch/Data No. Control of storage facility.	Reject. Retest and relife.
7. Pre-treatment of surfaces.	Requirements specified by (a) and (b) material properties: Cleaning, surface removal, dressing or chemical treatment	Use tested procedure (mistake proofed). Trained staff.	Re-treat.
8. Assembly.	a. Component fit-up: Correct components Location b. Application: Type Mix Quantity Temperature	Inspection. Use of jigs. Metering by calibrated dispenser. Use of tested procedure (mistake proofed). SPC. Trained staff.	Re-jig or select correct components. Reject or re-apply.

Activity	Quality Requirement	Method of Validation/Control	Corrective Action
9. Cure	Requirements specified by adhesive supplier: Time Temperature Pressure Heating/cooling rate	Use tested procedure (mistake proofed). Time/temperature records. SPC. Trained staff.	Reject or re-cure.
10. Final inspection	Joint meets design requirements: Strength Environment Appearance Reliability Durability	Test programme. Review of process documentation and records.	Reject. Concession. Design change.
11. Pre-use storage	Joint meets design requirements: Strength Environment Appearance Reliability Durability	Correct storage review of test reports, supplier information.	Reject. Design change.
12. Service	Joint meets design requirements: Strength Environment Appearance Reliability Durability	Service monitoring. Joint failure rate.	Repair. Design change.

Table 3.2.2 QUASIAT, Level 2, Sub-model 1

Activity	Quality Requirement	Method of Validation/Control	Corrective Action
7. Pre-treatment of surfaces.	Requirements specified by (a) and (b) material properties: Cleaning surface removal/ dressing chemical treatment	Use tested procedure. Monitor parameters with SPC.	Corrective action. Re-treat.
8. Assembly.	a. Component fit-up: Correct components, Location b. Application: Type Mix Quantity Temperature	Use of jigs. Metering by calibrated dispenser. Use of tested procedure (mistake proofed). SPC.	Re-jig or select correct components. Reject or re-apply.
9. Cure.	Requirements specified by adhesive supplier: Time, Temperature, Pressure, Heating or cooling rate.	Use tested procedure (mistake proofed). Monitor time/ temperature records with SPC.	Reject or re-cure.
10. Final inspection.	Joint meets design requirements: Strength Environment Appearance Reliability Durability	Test programme (samples) with SPC.	Reject. Concession design change.
11. Pre-use storage.	Joint meets design requirements: Strength Environment Appearance Reliability Durability	Correct storage review of test reports, supplier information.	Reject design change.
12. Service.	Joint meets design requirements: Strength Environment Appearance Reliability Durability	Service monitoring. Joint failure rate.	Repair design change.

Table 3.2.3 QUASIAT, Level 2, Sub-model 2

Activity	Quality Requirement	Method of Validation/Control	Corrective Action
7. Pre-treatment of surfaces.	Requirements specified by (a) and (b) material properties: Cleaning surface removal/ dressing chemical treatment	Use tested procedure (mistake proofed). Trained staff.	Re-treat.
8. Assembly.	a. Component fit-up: Correct components, Location b. Application: Type Mix Quantity Temperature	Use of jigs. Use tested procedure. Trained staff.	Re-jig or select correct components. Reject or re-apply.
9. Cure.	Requirements specified by adhesive supplier: Time Temperature Pressure Heating or cooling rate	Use tested procedure. Monitor time/ temperature records with SPC. Trained staff.	Reject or re-cure.
10. Final inspection.	Joint meets design requirements: Strength Environment Appearance Reliability Durability	Test programme. Review of process documentation and records.	Reject. Concession. Design change.
11. Pre-use storage.	Joint meets design requirements: Strength Environment Appearance Reliability Durability	Correct storage. Review of test reports, supplier information.	Reject. Design change.
12. Service	Joint meets design requirements: Strength Environment Appearance Reliability Durability	Service monitoring. Joint failure rate.	Repair. Design change.

Table 3.2.4 QUASIAT, Level 2, Sub-model 3

Activity	Quality Requirement	Method of Validation/Control	Corrective Action
7. Pre-treatment of surfaces.	Requirements specified by (a) and (b) material properties: Cleaning surface removal/ dressing chemical treatment	Use tested procedure (mistake proofed). Trained staff.	Corrective action. Re-treat.
8. Assembly.	a. Component fit-up: Correct components Location b. Application: Type Mix Quantity Temperature	Use of jigs. Metering by calibrated dispenser. Use of tested procedure (mistake proofed). SPC. Trained staff.	Re-jig or select correct components. Reject or re-apply.
9. Cure.	Requirements specified by adhesive supplier: Time Temperature Pressure Heating or cooling rate	Use tested procedure (mistake proofed). Trained staff.	Reject or re-cure.
10. Final inspection.	Joint meets design requirements: Strength Environment Appearance Reliability Durability	Test programme Review of process documentation and records. Trained staff.	Reject. Concession. Design change.
11. Pre-use storage.	Joint meets design requirements: Strength Environment Appearance Reliability Durability	Correct storage. Review of test reports, supplier information.	Reject. Design change.
12. Service	Joint meets design requirements: Strength Environment Appearance Reliability Durability	Service monitoring. Joint failure rate.	Repair. Design change.

Table 3.2.5 QUASIAT, Level 2, Sub-model 4

Activity	Quality Requirement	Method of Validation/Control	Corrective Action
7. Pre-treatment of surfaces.	Requirements specified by (a) and (b) material properties: Cleaning surface removal/ dressing chemical treatment	Use tested procedure (mistake proofed). Trained staff.	Re-treat.
8. Assembly.	a. Component fit-up: Correct components, Location b. Application: Type Mix Quantity Temperature	Use of jigs. Use tested procedure (mistake proofed). Trained staff.	Re-jig or select correct components. Reject or re-apply.
9. Cure.	Requirements specified by adhesive supplier: Time Temperature Pressure Heating or cooling rate	Use tested procedure (mistake proofed). Monitor time/ temperature records. Trained staff.	Reject or re-cure.
10. Final inspection.	Joint meets design requirements: Strength Environment Appearance Reliability Durability	Test programme. Review of process documentation and records. Trained staff.	Reject. Concession. Design change.
11. Pre-use storage.	Joint meets design requirements: Strength Environment Appearance Reliability Durability	Correct storage. Review of test reports, supplier information.	Reject. Design change.
12. Service	Joint meets design requirements: Strength Environment Appearance Reliability Durability	Service monitoring. Joint failure rate.	Repair. Design change.

4 QUALITY MANAGEMENT MODEL, PC BASED

4.1 Installation and Use of the PC-based Quality Model

It has been found helpful to document the model in terms of a software package. It is not essential, but in paper-based format the model is inevitably less compact. By the software route, all options can be presented and then customised by the conscious addition of all essential items; but only the relevant detailed texts need to be visible.

The QUASIAT package is a computer based generic quality model written in Microsoft Visual Basic for Windows (V3.0), with the following software and hardware requirements.

1. Any IBM compatible PC with an 80386 processor or higher
2. A hard disk (C-drive) - optional, but recommended
3. A 3½ " floppy drive
4. A SVGA display
5. One megabyte of memory
6. A mouse
7. Microsoft Windows version 3.0 or higher.

Visual Basic and Windows are trademarks of Microsoft Corporation.

To install the QUASIAT package in a PC hard disk insert the disk into drive A and type **a:\ setup.** Follow the setup program instructions as they appear on the screen. Select the directory where you want to install this package. The setup program installs the QUASIAT files in the selected directory. In addition to these files the setup program installs a few system files in the Windows\system directory of your PC. These files (VBRUN300.DLL and files with .VBX extension) are essential for running the QUASIAT package. If your PC is linked in a network, it may not allow QUASIAT installation to alter the network Windows directories. In this case, exit from the network and install the package via local Windows.

To run this package select **Run** from **File** in the Menu and type QUASIAT1.MAK at command line. As QUASIAT starts running a welcome message will be displayed on the screen. Follow instructions as they are displayed on the screen. QUASIAT is a self-structured package with its own menu which consists of five headings:

File

This includes **New Project, Open Project, Save Project** and **Exit. Open Project** loads the previously saved QUASIAT application. **Save Project** saves the current application in a selected file. A QUASIAT application example based on an actual adhesive bonding application can be found by clicking on **Open Project** and selecting the file *example.txt* in the QUASIAT directory. **Exit** ends the program. Before using the exit command the current application should be saved. Otherwise the user will lose all the data collected during working with this package.

Help

An introductory text file explains how this quality model has been structured and gives the users an overall view on how QUASIAT works. Should you like to read this introduction select **QUASIAT** from **Help** in the **Menu**. Select **Quit** from **Help** whenever you want to return to the main **Menu**.

Quality for Design

This includes all the steps in design of an adhesive bonded joint, such as **specification review, selection of materials, design, and verification of design**.

Quality for Manufacturing

All the steps in the manufacture of an adhesive bonded joint namely, **prefitting, surface preparation, assembly, cure, final inspection** and **storage and service** are documented. However, not every application employs all of these steps in their manufacturing process. At the beginning of the program a frame will be displayed in the centre of the screen asking a user to select those processing steps that are not used in their application. By highlighting these processing steps at the start, the program removes them automatically from the quality control steps. Should you select them incorrectly you have to run the program again, after you have exited from it.

Quality Plan

By clicking on **Quality Plan** the program automatically processes a quality plan based on the information given earlier by the user. The content of the quality plan can be saved as an ASCII file by selecting **SaveAS** from file in the **Quality Plan** Menu. This file can be edited by Notepad (Windows editor).

Remember to save the quality plan at this stage since **Save project** does not save the quality plan.

The two main analytical sections provide full coverage of all stages from the specification review to the despatch of the bonded components or product. Each sub-section can be accessed individually and comprises a series of questions and some Information Files. Generally, each topic starts with an abstract explaining, in general, the nature of activities that are associated with that section. Next, is a questionnaire with either a binary answer or a set of options to be selected. The quality model is based initially on the response of users in defining the quality requirements for each stage of bonding and, depending on the selected answer/options, the quality model displays a series of messages or advice. Whenever the required quality parameters are not defined, the program asks the user to open the available datafiles (Checklists) which can be used to prepare and define a new set of quality requirements. A feedback system takes the user back to the relevant stage where incomplete or incorrect inputs can be readdressed. As the user goes through the questionnaire and answers the questions, the program displays further details on each relevant stage of bonding and ultimately provides a recommended list of appropriate methods of validation and control for design and manufacturing activities.

Both the **Quality for Design** and **Quality for Manufacturing** sections are supplemented with Information Files that cover all the necessary requirements for adhesive bonding. Files associated with the **Quality for Design** section are constructed in a way that can be used as a data resource (information, see section 4.3) or a quality control method (checklists, see section 4.2 and APPENDIX B).

As in the text-based description of the model, the **Quality for Manufacturing** section is defined by the two key features that determine the approach to quality management of the production stages (Production Rate and Level of Mechanisation) and their four high/low combinations.

As a quality control method is applied to an activity, a corrective action may be needed to ensure that the required level of quality will be achieved. In many cases such a remedial action may require feedback to an earlier stage. Although all stages are accessible individually in this model, an automatic feedback system is invoked whenever it is necessary to allow the user to adjust his quality system in the light of new information. An information system also advises the user of capability and compatibility of control methods with requirements for each application.

Finally, the package sets up a data file where the required information will be collected automatically as the users respond to the set of questions that are specifically designed for this purpose at different stages. At the completion of the progression through the model, all required information is processed into a text file in the form of a **Quality Plan** table with the related information and instructions. The model allows the user to input specific information so that the generic template can be customised for any given particular bonding application.

4.2 Introduction to Checklist Documents

Two sets of blank checklists, on Specification Review and Design, are collected in APPENDIX B at the end of this book. If the QUASIAT software is loaded in a PC with Microsoft 'Word' available, then they can be accessed by opening "CHKLIST.DOC" file from this word processing environment.

Specification Review Checklists, is divided into three sub-sections:

Quality Requirements The first sub-section deals with the quality requirements which will generally be the subject of negotiation between a customer and a supplier. Having this information available helps both parties to put forward realistic and achievable requirements. Consequently, defining the requirements and specification for a product in detail helps manufacturers to achieve a high standard of quality and customer satisfaction. The information in this sub-section is formatted in the form of a tree-chart type table. A column is included which can be used to identify the selection of individual requirements.

Concerned Parties & Data Resources The second sub-section lists the relevant parties concerned with manufacturing and also shows from where these specifications are likely to be derived.

Specification Change Finally, when a modification of the original set of requirements or specifications is needed the third sub-section gives a list of possible involved parties and the requirements that such a corrective action should satisfy or should conform to.

Design Checklists, the second set of tables, covers all specifications and requirements needed for designing and manufacturing an adhesive bonded joint including joint features, materials used for bonding, process conditions, quality control (inspection), service and maintenance. Separate tables are presented to show detailed specifications for:

Design
Adherend Materials
Consumable Materials
Adhesive
Storage of Adhesive
Design of Process

4.3 Information Documents

4.3.1 Introduction

All the following sections are prints of text files displayed at various times during operation of the QUASIAT software. The introductory file **About QUASIAT** is effectively described elsewhere and is omitted from this text. Several complex diagrams included here are not in the text files.

4.3.2 Introductory files

4.3.2.1 Verification of design

The design process should produce the technical information for purchasing, production/operation, inspection, maintenance, etc., in a clear and comprehensive manner. This is usually in the form of a manufacturing specification set with rational values and tolerance limits. The objective of this section is to ensure that the design meets the requirements specified and that progress is being made towards the objectives. The design output is classified into six sub-sections:

1- Features of Joint,
2- Selection, Storage & Despatch of Materials (i.e. adhesive, adherend, consumables),
3- Production/ Operation Instructions,
4- Machinery & Equipment,
5- Quality Control & Inspection,
6- Storage, Service & Maintenance.

4.3.2.2 Prefitting

It may be possible that the parts to be bonded can be dry fitted together to ensure a close contact of the mating surfaces. If two or more parts do not fit prior to being bonded, they are not likely to fit well enough after bonding to produce a satisfactory joint. If a high production rate exists where a reproducible fit accuracy can be established, then prefitting can be omitted. Nevertheless, the following parameters are the major requirements for checking that fit is satisfactory:

1- Size,
2- Shape, Colour and Angle,
3- Tolerance limits,
4- Order of Assembly.

4.3.2.3 Surface preparation

Surface preparation and an appreciation of the required surface condition prior to bonding are two of the most important steps in the entire bonding process. The main purpose of surface treatment is to guarantee wetting of the treated surface by the adhesive and in some cases to introduce beneficial functional chemical groups on the surface. Consequently, a strong bond between adherend and adhesive is achievable. Surface treatment can modify the adherend's surface in four ways:

1- Modify physical structure,
2- Change the surface topography,
3- Change the chemical nature of surface,
4- Remove a weak boundary layer.

There are various degrees and many methods of surface preparation for adhesive bonding. They can range from simple dusting or blowing away of loose dirt to the complete removal of unwanted surface layers. Some methods may include the use of hand tools whereas others may be carried out automatically on complex equipment. The quality requirements may vary from treatment to treatment but all share an important requirement- "even treatment across the bonding area".

4.3.2.4 Assembly

For component assembly, there is a wide availability of different methods of adhesive bonding. Some of these can restrict the freedom of design options for both manufacturing process and end-product. The method of adhesive application or the way that joints are to be assembled can affect the selection of an acceptable adhesive for the process, the quality or performance of the product, and very often, assembly cost.

Critical to the ultimate choice are :

1- Size and shape of parts to be bonded,
2- Specific area to which adhesive is to be applied,
3- Number of assemblies to be produced,
4- The required production rate,
5- Viscosity or other characteristic of adhesive.

4.3.2.5 Cure

Adhesives are often classified by the way in which the bond is formed:

1- Pressure Sensitive,
2- Contact,
3- Chemical Curing,
4- Heat Curing/Setting,
5- Liquid Release (loss of solvent or water).

For pressure sensitive and contact adhesive systems, pressure is a necessary prerequisite to form a bond, whereas chemical curing, heat curing/setting and liquid release adhesives will probably cure faster if heat is applied. Due to the wide range of adhesives available, it is not possible to consider in any further detail the way in which curing can be facilitated or cure time reduced. What is important is a uniform distribution of heat and/or pressure across and along the bonding area. If in any doubt, the adhesive supplier should be able to provide all the necessary information relating to optimisation of cure.

Each adhesive group requires different combinations of conditions or processes to ensure the formation of a good bond. The main conditions or processes to consider are:

1- Heat (temperature, time, distribution and method of application),
2- Pressure (amplitude, distribution and method of application),
3- Environmental Influences (lack/availability of oxygen or moisture).

4.3.2.6 Final inspection

In spite of the fact that final inspection is expensive and is "after the event" it has been traditionally employed as the major means of quality control. It is desirable to replace completely or to diminish the role played by final inspection by using a systematic and thorough in-process quality system. Nevertheless, when it is required by a customer it should primarily ensure that the customer requirements and specifications are met. Therefore, Final Inspection ties in with processing activities and consequently a quality control method is required to make sure that this process does what it is designed to do. In this section implementation of a quality control method on Final Inspection will be discussed. Requirements for Final Inspection are generally defined by the customer and they are concerned with integrity and reliability of the bonded joint.

4.3.2.7 Pre-use storage and service

Accelerated curing is not always possible. For example, when the adherend's heat distortion temperature is low, application of heat may damage the component. During a period of Pre-use Storage it is crucial to make sure that the bonded joints are not exposed to damaging environmental or mechanical conditions. The main purposes of Pre-use Storage are to ensure the following.

1- that joints gain their ultimate strength,
2- that final products are despatched in the correct sequence.

In many adhesive bonding applications, in particular, when a joint is exposed to a harsh environment, regular in-service inspection of the joint could be part of the customer's requirements. In such cases design should not only consider ease of service and inspection of the joint but also should include a quality control method to ensure that the integrity of the joint is maintained at least until the next inspection and service or maintenance procedure has been carried out.

4.3.3 Information files

4.3.3.1 Materials selection

Materials required for adhesive bonding can be classified into three groups:

1- Adherends,
2- Consumables,
3- Adhesives.

Before making any selection, we advise you to look through this file. This file provides valuable advice on selection of materials and a checklist of the most common properties which have to be considered in most adhesive bonding applications. For detailed information use the checklist document or use design output.

Adherends

Whether the required materials to be bonded have already been defined by the customer or can be selected by the designer, it is important to have knowledge of their properties.

Trade Name and Supplier
Composition or Grade
Appearance (i.e. weight, shape, finish, colour, etc.)
General & Bonding Area Dimensions
Mechanical Properties
Physical Properties
Storage, Handling & Despatch

Consumable Materials

Consumable Materials can be classified into two groups; (a) pre-bonding, and (b) post-bonding applications. Quality of adhesive bonded joints depends on the quality of the bonding surfaces and consequently, on the quality of the consumable materials used for surface treatment. Therefore, it is, in many cases, vital to use consumable materials with correct properties. Use of appropriate consumable materials for post-bonding treatment is equally important. Use of inappropriate materials may damage the integrity of the bonded joint, particularly when the joint has not yet gained its full strength. The following table shows the major common features to be listed of consumable materials used in adhesive bonding.

Trade Name and Supplier
Composition & Grade
Form & Shape
Storage, Handling and Disposal.

Adhesive types

With the wide range of adhesives on the market and, in particular, with the general access to computerised selection systems, selection of an appropriate adhesive may not seem to be a difficult task. However, we recommend a thorough investigation into the design and operation processes prior to any selection. This package does not provide such a facility but a checklist of the necessary requirements and specifications is available in this system which gives the user an opportunity to narrow the choice of adhesive for the application.

All adhesive suppliers publish technical data, health and safety information and notes on the use of their products. However, the choice of the appropriate adhesive can be simplified if qualifying 'Ground Rules' are established first. Four basic factors should be considered when analysing an adhesive application.

1- Type of materials to be bonded is important.
2- All possible process or bonding procedures should be thoroughly investigated. The most simple process that satisfies the final requirements should be selected.
3- Before an adhesive can be selected, all service conditions must be considered.
4- The economics of the system must be evaluated.

An important step in the bonding process is adhesive storage. Requirements for the storage of adhesive to some extent depend on the shelf life, the volume of adhesive in use at the production line and its cost. The full cost of the adhesive could include the following.

1- The production waste, plus material used to join product.
2- Evaporation loss.
3- Excess adhesive application.
4- Amount of adhesive remaining in containers.
5- Solvents used to clean up production equipment.
6- Material loss due to contamination and unqualified products.

Adhesives can be classified in many different, but no optimum way. Most adhesives can be classified chemically into four types according to the setting mode, as follows.

Chemical Setting or Reactive
Liquid, paste or film, not reversible, thermosetting.

Solidification on Cooling
Hot melt or heat setting. Some are reactive, then no longer thermoplastic.

Loss of Liquid
Loss of solvent or water, by evaporation or diffusion. Immediate closure of porous substrates. Delayed cure, then 'contact' or pressure sensitive, or heat reactivated.

Pressure sensitive
Applied as solution, or hot melt, or in tape form

This classification is expanded with many (chemical) examples plus associated assembly features and physical form in Figure 4.3.3.1.

A user must also consider both the **performance properties** and the **manufacturing issues**. These engineering considerations are expanded in Figure 4.3.3.2.

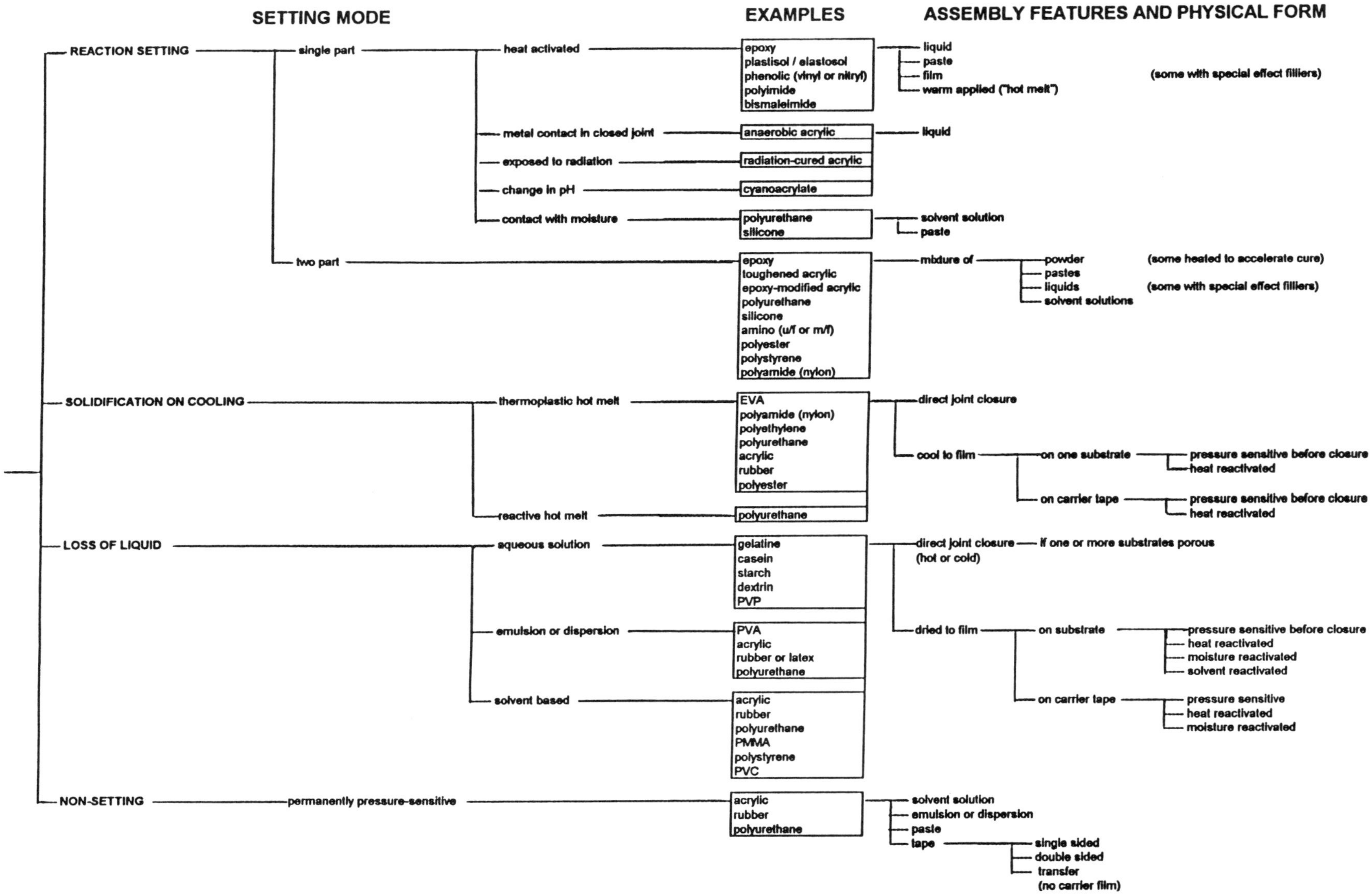

Fig. 4.3.3.1 Classification of adhesives, by setting mode, assembly features and physical form

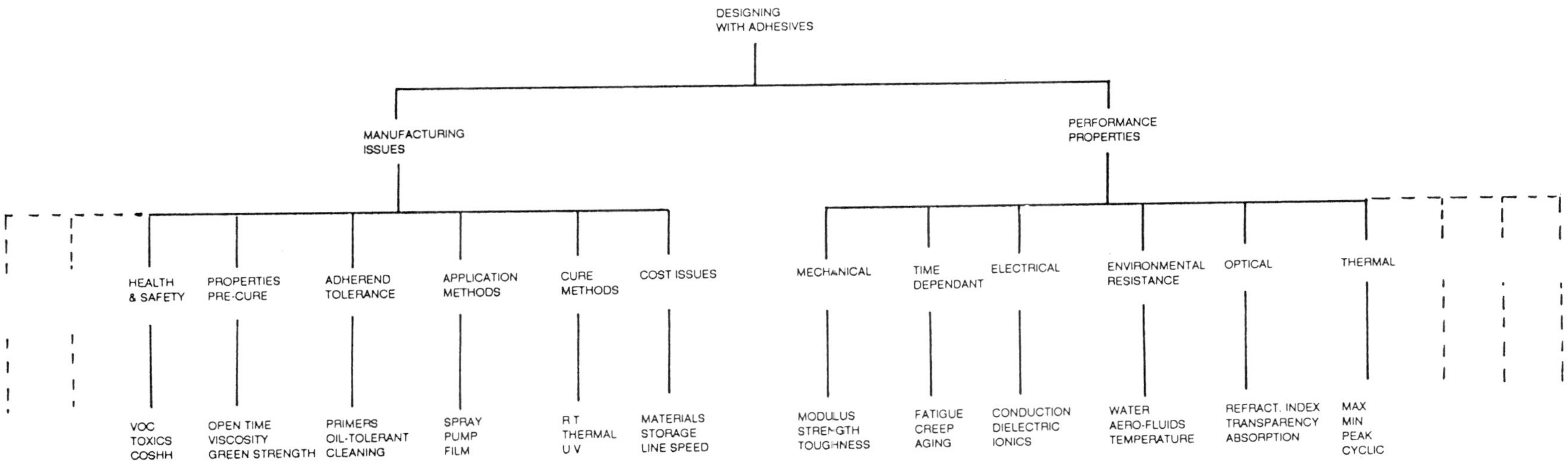

Fig.4.3.3.2 Performance and manufacturing issues when designing with adhesives

4.3.3.2 Design (recommendation on quality control methods)

In order to help users of QUASIAT in their decision to select an appropriate control method brief remarks on the quality control methods mentioned earlier (QFD, Taguchi, FMEA) are presented here.

QFD

The foundation of QFD is the belief that products should be designed to reflect customers' desires and tastes. Therefore this method is most suitable for applications that are based primarily on the detailed requirements and specifications of the customers and where inter-functional planning and communications among marketing people, design engineers and manufacturing staff are required.

Taguchi

The off-line Taguchi method defines quality control in relation to design for quality. This is a continuous quality improvement programme that includes continuous reduction in the variability of product performance characteristics about their target values. In this method statistically planned experiments can be used to identify the setting of product (process) parameters that reduce performance variation. Therefore, this method is less suitable for applications on which performance characteristics are not based on numerical measurements, but on an order categorical scale such as Poor, Fair, Good or Excellent.

FMEA

Failure Mode and Effect Analysis is an important method to identify all possible failures and their effect on the system. The objective is to classify failures according to their effect. FMEA provides an excellent basis for classification of characteristics. When engaged in FMEA, it is wise to bear in mind that the severity of the failure is not the only important factor. One must also consider the probability of failure. One purpose of FMEA is to direct the available resources toward the most promising opportunity for improvement. As far as adhesive bonding is concerned, this method offers a quality control technique that is most suitable for low rate production (i.e. highly cost sensitive production line) and where multiple processes are inter-linked.

DFMA

Design For Manufacturing and Assembly reflects manufacturing problems in an assembly line and how these problems should be handled. This technique is concerned with reducing the cost of a product through simplification of its design. In a multi-part assembly line the best way to achieve this cost reduction is:

1- to reduce the number of individual parts that must be assembled,
2- to ensure that the simplest process route has been selected.

This technique has been successfully applied to high and low production rate, multi-part assembly lines. It is most suitable for those adhesive bonding applications that consist of multiple joint assemblies.

4.3.3.3 Design verification (methods of control)

Five control methods suggested in QUASIAT for verification of design are discussed here.

What is the qualification test?

There are two types of qualification test:

1- Destructive Test
2- Non-destructive Test

Based on type and geometry of the joint, numerous mechanical testing techniques are available in the technical literature on adhesive bonding and are also well documented in ASTM (Vol.3) or BS 5350.

The following are some common Non Destructive Tests:

1- Sonic, Ultrasonic, Visual and Dielectric Testing,
2- Thermal Infra-Red Inspection,
3- Core and Proof Sampling (not entirely ND).

On what parameters can a design calculation be based?

There are various base line parameters that an alternative calculation can be based on. Some of these alternative parameters are presented here:

1- Optimum thickness of adherend,
2- Optimum overlap length,
3- Mean fracture stress,
4- Bondline thickness,
5- Alternative joint design.

What is the comparison method?

As joints have been assembled in a quality controlled environment, data will be available on the expected level of reliability of the joint. Application of the comparison method for design verification, relies basically on two principles.

1- Using past practice as a standard, and
2- Using revised past practice as a standard.

Nowadays, comparison methods are widely used for verification of design. In particular, application of statistical methods, i.e. Bye's Theorem, which is based on previous data, provides a wider coverage and a more reliable result.

What is Design Review?

Design reviews are sometimes made mandatory by customer specification. To conduct it, specialists who are not directly associated with the development of the design should be part of the review team. Design review should be conducted in a formal manner and should cover all quality-related parameters. Design reviews can be conducted at several stages during the development of the design. The following table represents the major activities in a comprehensive design review:

Activities	Stages*		
Review of "Design, Test & Calculation" data.	P	I	F
Review of "Design for optimum reliability".	P	I	F
Review of "Inspection, Test & Control" procedures to ensure that they will be carried out efficiently.		I	F
Review of "Production Processes", including production rate, production line sequence, automation & schedules and cost effectiveness.		I	F
Review of "Post-production Processes", including installation, maintenance, control and customer special requirements.		I	F
Review of "Materials used in Production Line". To ensure that the materials selected will perform as required. To ensure that the accepted parts & materials are available and will meet cost & delivery schedules.		I	
Review of "Machinery, Equipment & Tools" required in production line. To ensure that they are capable of what they are required to do, cost effective and safe to employ.		I	
Review of "Storage & Despatch Processes". To ensure that the product is capable of being handled without damage.		I	

*** P = Preliminary Design Review**
*** I = Intermediate Design Review**
*** F = Final Design Review**

Clearly, for a simpler situation, a review can be carried out in a relatively simple way.

4.3.3.4 Design verification (corrective action)

To carry out a corrective action first it is necessary to define responsibilities and activities clearly. The following are the major responsibilities of a corrective action group, which may well be the original design team and the topics they must ensure.

1- The implementation of the design changes complies with specifications and requirements of the original design.
2- The implementation of the design changes are cost effective and with minimum of schedule changes.
3- The design changes are implemented correctly through-out all stages.
4- The customer agrees with the changes.

The corrective action group discusses the various methods by which the changes can be carried out. Therefore, activities of the corrective action group are based on the following requirements:

Does the design require............ (a) Minor Change
(b) Major Change
(c) Total Reconstruction ?

How will the design changes be introduced ?

Who authorises the changes ?

Who takes responsibility to make sure that the changes will be implemented correctly and effectively ?

Does the customer agree with the changes ?

4.3.3.5 Design validation

Design validation is a new requirement of ISO 9000 and its objective is to ensure that the product itself complies with the design intent. This would normally mean some product testing if only on a prototype basis.

4.3.3.6 Surface preparation

There is a wide range of surface treatments available. Many adhesive or adherend suppliers give advice and best practice is collected in Standards, currently under revision; ISO 4588 (for metals) and ISO 13895 (for plastics). They can be classified according to the nature of treatment into five groups:

Cleaning/Degreasing
Removal of loose solids can be accomplished with a clean brush or blast of clean, dry air. Organic solvent or alkaline aqueous solution, removes organic materials such as grease, oil and wax from adherend surface. This can be accomplished by wiping, dipping or spraying.

Surface Roughening
Techniques where abrasive materials are employed to remove unwanted layers and generate a roughened surface texture.

Chemical Treatments
Immersion of the adherend in an active solution which has the power to etch or dissolve a part of the adherend surface or change it in such a way that the treated surface becomes chemically active. An electrochemical reaction can also be included where current is transferred through an electrolyte between an auxiliary electrode and the adherend surface, e.g. anodising and this introduces several additional factors to be controlled.

Physical Treatments
Techniques where the adherend surface is cleaned and chemically modified by exposure to excited charges or species. Techniques such as Corona Discharge, Plasma, Flame or Ultraviolet/ozone are examples in this group.

Primers
Alternative surface treatments, often simpler than chemical or physical methods, applied by dipping, brush or spray. They can chemically alter the surface (e.g. silane coupling agents, chromate conversion coatings), or protect the preferred surface already generated by another method (e.g. aerospace primers after anodising).

The selection of a pretreatment will be made on the basis of issues such as cost, production, durability or health and safety. Pretreatment facilities can include equipment, chemicals and consumables. Key surface features which should be kept in mind are wettability, roughness, soundness, stability, contamination-free, uniformity and adhesive compatibility.

4.3.3.7 Assembly (application of adhesive)

The following are the aspects of the application of adhesives where quality control actions are relevant:

1- Preparation of adhesive, including mixing and heating.
2- Pot-life vs. Volume of the adhesive which is going to be used.
3- Dispensing Techniques, tools and applicators.
4- Dispensing appropriate amount of adhesive evenly on the bonding channel/area.
5- Repeatability of dispensing.
6- Maintenance of the dispensing equipment.
7- Health and Safety requirements.

A comprehensive diagram from IVF summarises all the options in Figure 4.3.3.3.

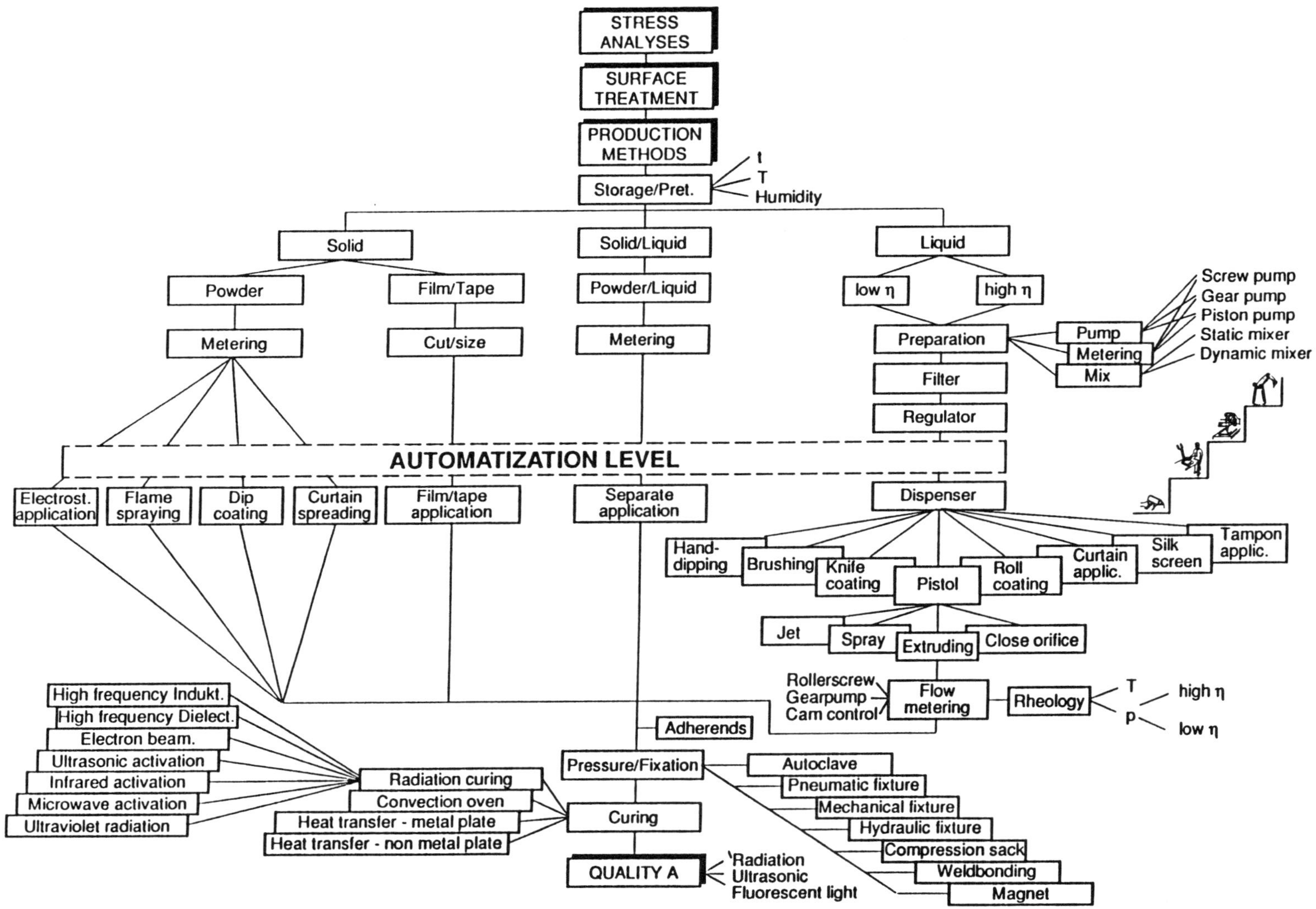

Fig. 4.3.3.3 Dispensing and curing of adhesive types (courtesy of IVF)

4.3.3.8 Assembly (bonding)

These are the major aspects of bonding which affect quality:

1- Assembly Methods.
2- Component position.
3- Sequence of assembly in a multiple joint.
4- Bonding pressure, use of appropriate fixtures, or self-locating.
5- Post-bonding treatment.
6- Handling of the assembled component.
7- Maintenance of the assembly equipment.

4.3.3.9 Assembly (corrective action)

When a quality control system is applied to an activity or series of activities a corrective action may be required. These actions are taken to readjust the system:

1- **Design Change**: This action refers to change in processing activities, use of different tools or equipment, new set of specifications (tolerance limits) or implementation of a different quality control method.

2- **Rework**: This is for bonding applications where either the adhesive is applied incorrectly or the bonding parts are misplaced. It is necessary to work out how to dissemble the bonded parts and how to remove the applied adhesive from bonding area. Surfaces of the reworked joint may need to be treated. Cost efficiency is an important factor in selection of this action.

3- **Reject**: This should be considered as a last resort. Rejection may not imply a total waste if concessionary use of the component/product or downgrading is possible.

4.4 Conditions of Use

The QUASIAT generic quality management model for adhesive bonding was developed by Cranfield University for the Centre for Adhesive Technology, TWI.

The software [Version 1.2 of June 1997] is copyright material to TWI and is supplied on the terms set out below.

1. TWI warrants its title to and property in the software is unencumbered and that it has the right and authority to license the same.

2. Subject to the above, all conditions, warranties, terms and conditions expressed or implied statutory or otherwise in respect of the software are hereby excluded.

3. All work undertaken by TWI and its sub-contractors is experimental in its nature. Computer programs are developed and written using sound scientific principles. Whilst every effort has been made to ensure the contents of the software are correct, it is issued on the condition that it is intended only as a guide to the most appropriate quality management procedures for any adhesive bonding application and there is no guarantee that the software is error-free. It is supplied with no further technical support.

4. The software is supplied on the basis that the user will verify any results obtained to ensure, so far as is reasonably practicable, that the proper use of such results will be safe and without risk to health.

5. TWI and Cranfield University accept no liability for damages incurred in the use, failure to use or misuse of the software or the interpretation or misinterpretation of results derived from the software.

6. TWI and Cranfield University are unable to guarantee any design or specification derived from the software and accept no liability whatsoever for any damages incurred as a result of the use or publication of such designs or specifications, nor for any systems or products resulting from the software.

7. Other than as is provided in the above conditions, the user accepts all risks and liability for the use of the software.

<u>Acceptance of this software will be deemed to imply acceptance of the foregoing terms in full.</u>

5 SOME EXAMPLES OF THE USE OF QUALITY TOOLS AND TECHNIQUES, APPLIED TO ADHESIVE TECHNOLOGY

5.1 Introduction

Examples of most of the Tools and Techniques for Quality Management are given in the excellent SMMT booklet of the same name, referenced in section 6.1. This chapter contains worked examples of several, applied in realistic, industrial, adhesive bonding applications. All the examples described will be valuable to participants in simultaneous engineering teams, when 'pre-production quality assurance' (according to Nissan), or 'off-line quality control engineering' (according to DTI's Managing in the '90s) are used to resolve product or process issues prior to production, to reduce development time and encourage lean manufacturing.

The Wide Choice of Quality Tools and Techniques (courtesy of IVF)

5.2 Quality Function Deployment (QFD) - Automotive Headlamp Assembly, Carello Lighting

QFD is an advanced quality planning technique. It is a means of translating customer needs into the appropriate technical requirements at every stage of product development and production. It requires significant effort to carry it through to a conclusion, but it is a valuable method of logically collecting all the necessary requirements and then communicating them to the project team in a clear and standard format. The method is summarised below in Figure 5.2.1 and 5.2.2.

In this example QFD is used to optimise important features of an adhesively bonded automotive headlamp, in particular the design of the bonded joint, materials and processing details. Details of the assembly process are tabulated in APPENDIX A.

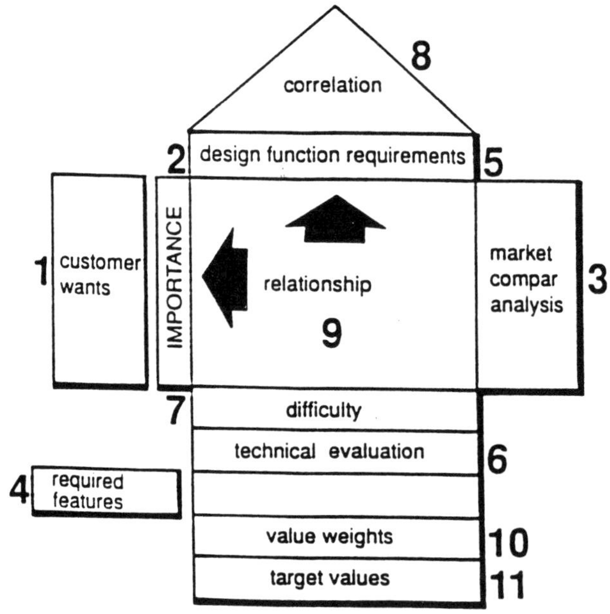

The team 1 brainstorms customer wants
2 determines the relative importance of these wants
3 evaluates market competitiveness
4 identifies required standards
5 generates design requirements to meet customer wants
6 evaluates technical competitiveness
7 determines degree of difficulty of requirements
8 identifies how well the requirements relate to customer wants
9 determines how well the requirements relate to customer wants
10 prioritizes by value weighting the requirements to satisfy the customer better than the competition

Fig. 5.2.1 The QFD product planning matrix (courtesy of SMMT)

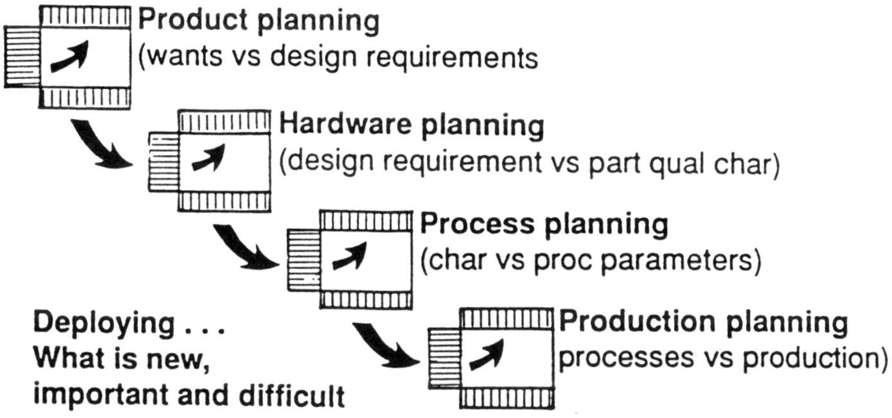

Fig. 5.2.2 The QFD process (courtesy of SMMT)

Building the house

The foundation of the house of quality is the customer's needs and requirements. Building the house starts with defining these requirements.

What do customers want?
The customer's requirements, called customer attributes (**CAs**), are phrases that customers use to describe products, or product characteristics. A typical list of such attributes for a bonded headlamp is shown in Table 5.2.1. These primary attributes can be sub-divided to provide a clearer definition of the fundamental features. In this example, the characteristics that represent the customer's requirements in relation to the adhesive bonded joint (body to lens) have been selected, i.e. **No Leaking** , **Durable Joint** and **Ease of Repair**. These attributes are placed in rows in the far left column of the house (Figure 5.2.3).

Are all CAs equally important?
Rarely can a design solution be found to satisfy all needs. Usually, designers have to trade off one benefit against another. To bring the customer's viewpoint into the design process, **CAs** are weighted according to team members' direct experience. Table 5.2.1 shows such a weighting in the column 'Relative Importance' (RI).

Table 5.2.1 Customer attributes for a headlamp

Primary	Secondary	Tertiary	R.I %	Customer Perception 1	2	3	4	5
Appropriate lighting	Light	L - Lucas 197 - 5 - 1/2/3	17				# +	
		B - high strength	10			#	+	
	Strength	B - light weight	7				# +	
Strong		B - black / dark grey	1.5			+	#	
		B - high distortion temp.	4.5			# +		
		J - strong joint	5			#		+
Correctly fitted lenses	Insulation	J - no leaking	16			+	#	
		J - durable	18			# +		
Maintenance	Service	J - ease of repair	6			#		+
		B - ease of bulb fitting	5					# +
L = Lenses		R.I = Relative Importance	100	# A's product				
B = Body		J = Joint		+ Our product				

Will delivering the perceived needs yield a competitive advantage?
Manufacturers who aim to match or exceed their competition must first know where they stand in relative terms. Therefore, on the right side of the house, opposite the **CAs**, the customer perceptions of competitive product matched to our own should be listed. Comparison with the competition can identify opportunities for improvement. When the ratings for **No Leaking** are compared, we are in an inferior position. This problem has to be tackled if we want to be competitive. However, for the **Joint Durability**, our product is no different from that of the competitor. An advantage can be gained here by improving the durability of our product. Comparing the **Ease of Repair** perceptions, we have a very strong position which should be maintained.

How can a product be improved?
So far, we have been informed of customers' desires and the market situation. The customers tell us what they want, the market position tells us what to do against competition and the engineering knowledge tells us how to do it. The first house of quality, **House of Design** maps the relationship between these domains and to do this we must describe the product in engineering terms.

Along the top of the house of design, the design team lists those engineering characteristics, **ECs**, that are likely to affect one or more of the customer attributes. Engineering characteristics should describe the product in measurable terms and should directly affect customer perceptions. The **ECs** defined to address the customer's requirements in this example are as follows.

1- Tolerance in the dimension of the bonding channel width
2- Tolerance in the dimension of the bonding channel depth
3- Bead weight of sealant
4- Tolerance in the bead weight of sealant
5- The angle of the edge of the moulding
6- Durability of the sealant (exposed to oil, water, heat, vibration)
7- Strength of the sealant, measured as peeling strength

What influences can be linked with customers' perceived qualities?
When the engineering parameters are defined the interfunctional team should fill in the body of the house. This 'relationship matrix' represents the degree of influence that **ECs** have on **CAs**. Based on engineering expertise, customer responses and data from statistical studies or controlled experiments, the working party will be able to fill in the body of the house with reasonable confidence. This can be done using numbers or symbols to establish the strength of these relationships. In this example six abbreviations are used to indicate the following relationships.

SP = Strong Positive **SN = Strong Negative**
MP = Medium Positive **MN = Medium Negative**
WP = Weak Positive **WN = Weak Negative**

A change to one engineering parameter generally requires some alteration in the other parameters. In the roof of the house the correlation between **ECs** can be displayed. The correlation section of the matrix helps the designer to balance the engineering parameters with the customer perceptions. For example, to improve the integrity of our product against leaks, the tolerance of the bonding channel width can be reduced. However, reduction of bonding channel width tolerance has a strong but negative relationship with lens dimension, which in turn could jeopardise our strong position with respect to the strength of the joint.

Using the House

Once the core of the house is filled, it can be used to help set the targets which will be displayed on the bottom row of the house. To set targets for **ECs** realistically, first we should review our market position and how customer perception can be changed in our favour.

With respect to **No Leaking**, there are two **ECs** which are strong positive influences. Despite restrictions on reducing the bonding channel width, due to its negative effect on several other parameters, there is room for improving the current value of 8.0 to 9.0 mm without incurring significant disadvantage on the other **ECs**. The next step is to define the technical difficulty in achieving such a target, with the degree of difficulty numerically scaled from 1 to 5. Then an assessment is made, based on team judgement and available data, of the value or importance of such an improvement. Finally this alteration may incur extra cost and the cost associated with such an alteration should be estimated.

Now that a new target has been set for one **EC**, other targets can be defined in the same manner. Key points in setting these targets are as follows.

1 Internal relationship between **ECs** and **CAs**.
2 The **EC** which is most capable of turning the customer perception in our favour should be defined.
3 The degree of alteration of an **EC** depends on both positive and negative effects that it may have on the other **ECs** and consequently on customer perception.
4 The setting of realistic degrees of importance for each **EC**.

With that, the **House of Design** is completed, but the design review is not complete until the new targets have been correlated with the characteristics of materials to be used and the processing capabilities, i.e the cascade of QFD houses as shown in Figure 5.2.2.

The second and third houses of quality, the **House of Materials** and the **House of Processing** can be prepared in the same way and **ECs** of one house become the CAs in the next house. Engineering characteristics are now on the left side of the house, along with their relative importance from the previous house (Imputed Importance values defined in the **House of Design**) and they are correlated with material characteristics (**MCs**) along the top row. Then the body of the house is filled in, targets are set again, considering customer perception and the internal influence of the **MCs** on the **ECs**. Figures 5.2.4 and 5.2.5 represent the **House of Materials** and the **House of Processing** for fabrication of the headlamp.

The major benefit of QFD is to encourage the product development team to think together and in the appropriate direction. In this study, the customer perceptions of our product in terms of the **No leaking** and **Durability** characteristics have to be improved by setting new design parameters. The **House of Design** tells us how this can be achieved, i.e. by reducing the bonding channel width or by changing the angle of the bonding channel edge from 90° to 45°. In order to achieve these targets, materials with new specifications may be required. The **House of Materials** shows what kind of materials are required to satisfy these new design targets, i.e. the percentages of filler in the composite body and additives in the adhesive are defined. Finally, the **House of Processing** shows what alteration in processing stages are required, if a new set of materials is to be used, i.e changes in the nozzle orifice diameter, in the heating required to keep nozzle temperature high and the reduced pump pressure that can be used with a higher viscosity adhesive.

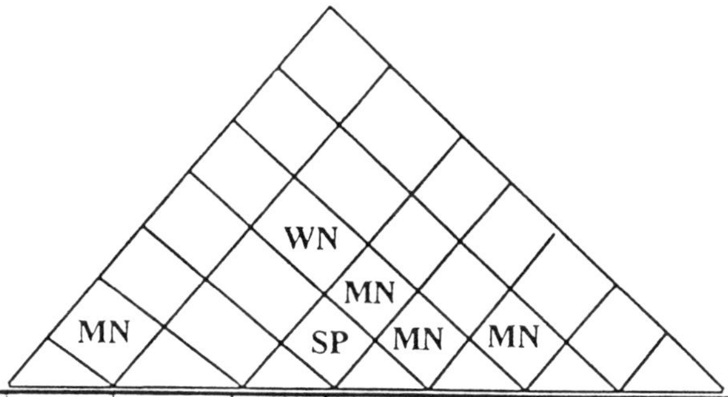

	R.I	CW	CD	BW	BT	AE	DS	PS	Customer Perception				
	%	-	+	+	-	-	+	-	1	2	3	4	5
No leaking	16	SP			MP			WN			+	#	
Durable joint	18			MP	MP	SP	SP	WN			# +		
Strong joint	5	MP	MP	SP	MP		SP				#		+
Ease or repair	6							SP			#	+	
Measurement Units		mm	mm	g	g	deg.	scale	MPa					
Our product (+)		9.0	6.5	29	±3.0	90	p	4			–		
A's product (#)		8.0	6.0	28	±2.5	90	g	6					
Technical difficulty		1	1	0	2	1	2	0	In scale of 1- 5				
Imputed Importance		8	6	6	4	4	17	5	%				
Estimated Cost		-0.04	0.04	0.0	0.1	0.01	0.1	0.0	% increase cost				
Targets		8.0	6.8	29	±2.5	45	g	4					

Key for the House of Design

Scale	Resistance (measured by reduction in shear strength after 200h exposure in the various conditions)	Abbreviation
Excellent	Less than 4 %	e
Good	Between 4 to 10 %	g
Moderate	Between 10 to 18 %	m
Fair	Between 18 to 28 %	f –
Poor	Over 28 %	p
Relative Importance		R.I
Bonding Channel Width		CW
Bonding Channel Depth		CD
Bead Weight		BW
Bead Tolerance		BT
Angular Edge		AE
Check Durability of the Sealant		DS
Peeling Strength		PS

Figure 5.2.3 QFD, The House of Design

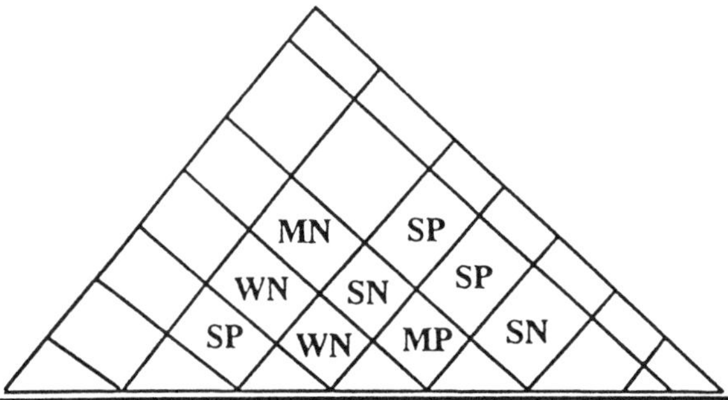

	R.I %	Filler addit-ive(B)	Viscos-ity	Wax addi-tive	PA	Melt point	Gap filling		Customer Perception 1 2 3 4 5
CW	8	MP					SP		+ #
CD	8								+#
Bead Weight	6					WN			+#
Bead Tolerance	3		SP	WP	MP	MN	MP		+#
Angular Edge	4	WP					WN		+#
Durability of Sealant	4			SP	MP		MP		+ #
Peeling Strength	5			MN					# +
Measurement Units		%	KmPa.s	%	%	°C	mm		
Our product (+)		40	22-30 E3	4.0	68.0	170	0.4-0.8		
A's product (#)		30	26-33E3	6.0	70.0	175	0.8-1.0		
Technical difficulty		1	1	1	1	1	0		In scale of 1- 5
Imputed Importance		5	8	7	6	10	8		%
Estimated Cost		-0.07	0.08	-0.02	0.05	0.1	0.0		% increase cost
Targets		35	24-32E3	4.5	70	175	0.4-0.8		

Key for the House of Materials

Parameters	Abbreviation
Polymer Additive	PA
Bonding Channel Width	CW
Bonding Channel Depth	CD

Figure 5.2.4 QFD, The House of Materials

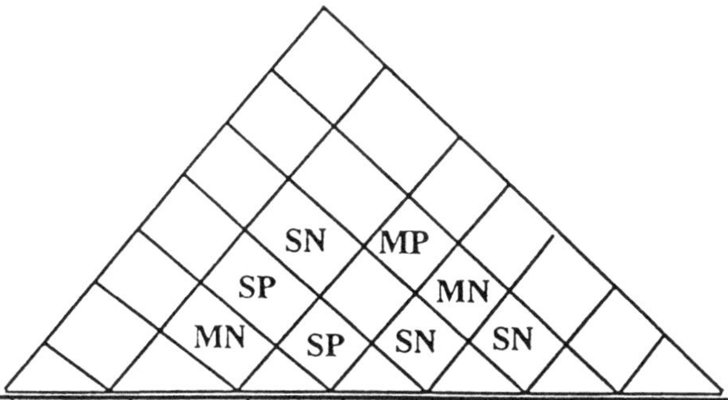

	R.I %	MP	NT	PP	DR	NO	NH		Customer Perception 1 2 3 4 5
Filler Additive (B)	5	SN							+ #
Viscosity (A)	8			SP	SP	MN	SP		
Wax Additive (A)	7				SP	MN	MP		+ #
Polymer Additive(A)	6				MN	WP			
Melting Point	10		SP	MN	WN		MP		
Gap filling –	8		MP				MN		
Measurement Units		Ton	^{0}C	Bar	g/Min	mm	^{0}C		
Our product (+)		7.2	146	4.6	650	3	25		
A's product (#)		6.1	150	4.4	700	3	60		
Technical difficulty		1	1	1	2	2.	2		In scale of 1- 5
Imputed Importance		10	2	4	5	4	16		%
Estimated Cost		-0.01	0.0	-0.02	0.01	0.03	0.07		% increase cost
Targets		6.8	146	4.4	700	2.5	80		

Key to the House of Processing

Process	**Abbreviation**
Moulding Pressure	MP
Nozzle Temperature	NT
Pump Pressure	PP
Dispensing Rate	DR
Nozzle Orifice Diameter	NO
Nozzle Heated Temperature	NH

Figure 5.5.5 QFD, The House of Processing

5.3 Taguchi (Design of Experiments) - Hard Disk Drive Actuator, IBM UK

Terry Twine, site statistician and **Martin Hall**, product scientist, IBM Havant (now Xyratex)

5.3.1 Introduction

In developing a product that incorporates adhesive bonds, much effort is often applied in the selection of the adhesive and surface treatment for the adherends, and pilot build of the product is normally conducted in a laboratory or under close engineering control. The bonding process would be determined by either recommendation from the adhesive manufacturer, previous experience or a combination of both and, if successful during product development, would be documented as it is for the manufacturing process. However, little attention is often paid to understanding the critical variables in the bonding process and problems frequently occur when the process is stressed during full-scale production.

At IBM Havant, a Design of Experiments (DOE) philosophy that incorporates "Taguchi" experimental methods has been widely used for quality improvement, particularly in the field of process optimisation. The "Taguchi" approach efficiently combines the knowledge of process operators, engineers and statisticians into a methodology that can evaluate the effect of manufacturing process variables on product performance. This paper will use a case study to illustrate the application of "Taguchi" methods in optimising an adhesive bonding process.

There has been much interest in the DOE/"Taguchi" methodology. This example, and several others, of Taguchi statistical design of experiments, applied to adhesive technology, from different sectors of industry using adhesives, will be available early in 1996, as a result of work carried out by the National Physical Laboratory and IBM UK, Havant Site. The work was funded by the Department of Trade and Industry, as part of a Measurement, Technology and Standards programme on the Performance of Adhesive Joints. The IBM assembly process is documented in APPENDIX A.

This chapter will be published as a paper in a special issue of the journal Materials and Design, dedicated to adhesive technology and it is included in this report with the permission of the authors.

5.3.2 Case Study

The case study used in this paper relates to an adhesive bonding process used in the assembly of an actuator for a hard disk drive. This actuator is a linear type where high quality bearings on a carriage ride along very smooth rails to accurately position the read/write heads over the disks. At the heart of the actuator is the core/rail assembly in which fine grain zirconia rails are bonded to a nickel plated sintered iron core using a single part anaerobic adhesive. The unit and the parts of interest are shown in Figure 5.3.1.

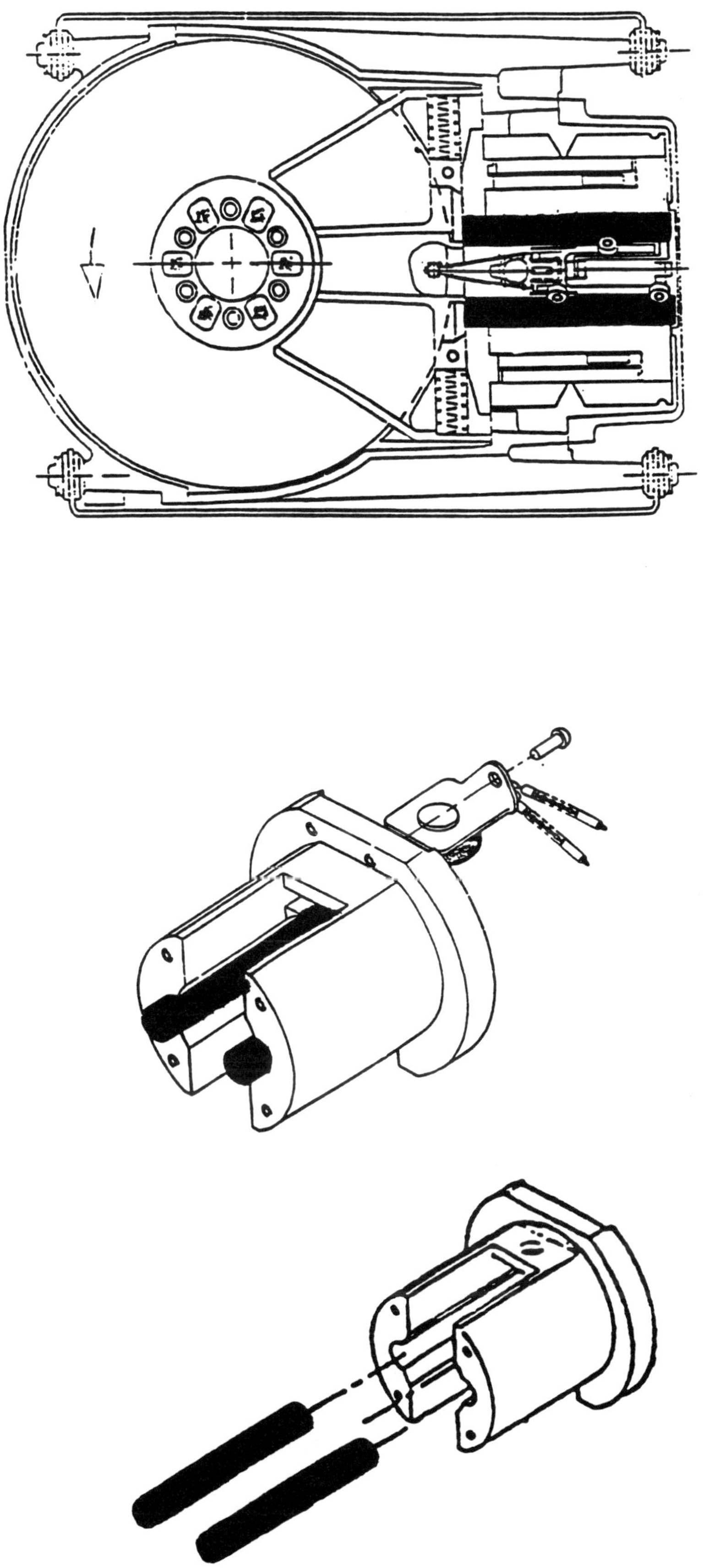

Figure 5.3.1 **Hard disk drive actuator, core to rail assembly**

During the early stage of product development as much as 50% failure rate of this bond was experienced during higher level assembly downstream in the actuator manufacturing process. Immediate action was taken by instigating a proof load test screen to yield core/rail assemblies of usable quality. However, the adhesive bonding process was clearly unreliable and coincidental action was initiated to define a reliable and robust process using DOE/ "Taguchi" methods.

5.3.3 Methodology

A key feature of the DOE/ "Taguchi" methods approach is the structure applied to the examination of a problem. The structure used at IBM Havant and applied to this problem was as follows.

5.3.4 Problem definition

The current core/rail bond does not meet the minimum strength requirement of 2kN which was established as the proof load test.

5.3.5 Objectives of the experiment

i) To define an adhesive bonding process that consistently delivers a core/rail bond strength in excess of 2kN.

ii) To define a cleaning stage in the adhesive bonding process that will remove any excess adhesive from the rail surfaces that contact the carriage bearings.

5.3.6 Product measurement system

i) Use a tensile tester to determine the axial load required to break the joint.

ii) Use a visual standard to identify any excess adhesive on the rails.

5.3.7 Process definition and causal factor establishment

This is the most critical stage in the approach as any experimentation to determine the optimum bonding process will depend on the correct identification of the key causal factors that lead to variation in the quality of the finished product.

Two major steps are necessary to complete this stage, firstly, a true definition of the process and secondly, the establishment of the key causal factors. Both steps must include all those involved in the process including line operators, process engineers and managers. To ensure that all views and ideas are identified it is often necessary to use brainstorming techniques, preferably facilitated independently. It is important to limit the number of factors so that the experimental stage is not cumbersome and costly. Again a facilitator can be very useful in getting consensus about identifying the key factors to be taken forward for experimentation, now referred to as control factors (voting is often accepted as a fair method for ranking the factors).

The core/rail bonding process and the key control factors identified for experimentation are shown in Figure 5.3.2.

5.3.8 Setting control factor levels and noise factor assessment

The levels of the control factors in the experiments are set to test the "robustness" of the process to each parameter. Table 5.3.1 shows the control factor levels chosen from the core/rail bonding process. Additionally, any unchangeable noise factors believed to be potentially influential should be identified and, if possible, incorporated into the experimental plan and subsequently analysed for significance. In the case of the core/rail bond, there are two rails in each core, and as the rail position was an unchangeable feature of the part, this was considered to be a noise factor which should be incorporated into the experimental plan.

5.3.9 Experimental design

The "Taguchi" method uses orthogonal arrays to define the experimental runs required to investigate the effect of the control factors. The size of the array is determined through an examination of the number of main effects and defined interactions which need to be studied. In this case an L(12) design was adopted as the objective was to investigate the main effects only. It should be noted that this L(12) array, as shown in Table 5.3.2, requires 12 experimental runs to yield highly significant information about an optimum process versus the 128 runs required for a full matrix experiment that would contain the optimum process combination of factors/levels.

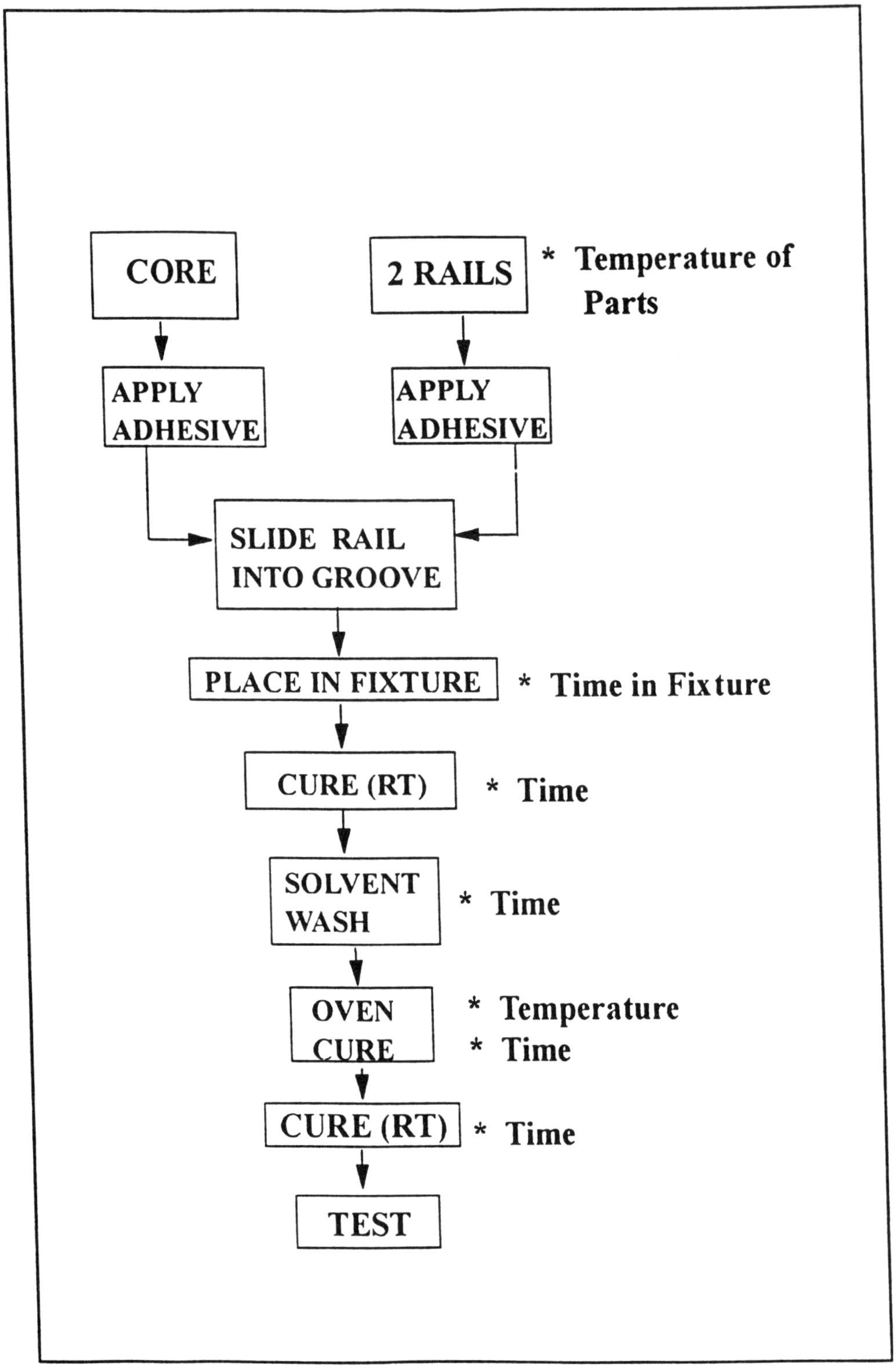

Fig.5.3.2 Core/rail bonding process and control factors (*)

Table 5.3.1 Control factor levels set for experimentation

CONTROL FACTORS - LEVELS			
LABEL	**FACTOR**	**LEVEL**	
		1	**2**
A	Temperature Parts	20 C	50 C
B	Time in Fixture	5 min	15 min
C	Time to Cure	30 min	120 min
D	Time in Oven	30 min	60 min
E	Time in Solvent	5 min	20 min
F	Time Cure Post Oven	4 hr	12 hr
G	Temp Oven Cure	20 C	100 C

Table 5.3.2 An L(12) orthogonal array

L(12) 2(11)											
No.	**1**	**2**	**3**	**4**	**5**	**6**	**7**	**8**	**9**	**10**	**11**
1	1	1	1	1	1	1	1	1	1	1	1
2	1	1	1	1	1	2	2	2	2	2	2
3	1	1	2	2	2	1	1	1	2	2	2
4	1	2	1	2	2	1	2	2	1	1	2
5	1	2	2	1	2	2	1	2	1	2	1
6	1	2	2	2	1	2	2	1	2	1	1
7	2	1	2	2	1	1	2	2	1	2	1
8	2	1	2	1	2	2	2	1	1	1	2
9	2	1	1	2	2	2	1	2	2	1	1
10	2	2	2	1	1	1	1	2	2	1	2
11	2	2	1	2	1	2	1	1	1	2	2
12	2	2	1	1	2	1	2	1	2	2	1
Group		1						2			

5.3.10 Sample sizes and replications

A realistic sample size is determined by consideration of the significance of the standard deviation of the response measurement and estimation of the differences between the responses achieved at the two control factor levels. In the case of the core/rail bond, an estimate was made that fifteen results at each of the two levels (i.e. thirty results in total) would be significant. The L(12) array would produce twenty four results in total due to each core having two rails. Therefore it was decided to duplicate the whole experiment and produce forty eight results in total which would allow for any inaccuracies made on standard deviation and significant difference.

5.3.11 Experimentation

The experiment was defined using the control factors/levels determined earlier and the L(12) array. To avoid any spurious effects from the experimentation, the sequence of runs was randomised and a different sequence was used in the duplicate experiment. The experimental plan with run sequence and results are shown in Table 5.3.3. It should be noted that the solvent washing produced adequate cleanliness of the rails at either level.

5.3.12 Analysis

The methods of analysis adopted for a "Taguchi" study are based on response graphs and the table of response means at different levels, see Figures 5.3.3 and Table 5.3.4 respectively. Analysis of Variance (ANOVA) was adopted to give statistical rigour to the judgements of significance, see Table 5.3.5.

As the experiment was designed as a noise array (ie. using the bond strength data from both rails in the assembly which were considered as being equal within the noise band of the bond strength values) with replication, the analysis can be based on raw data only and also with consideration of the "Taguchi" S/N characteristic. For the core/rail bonding experiment, the response measure of bond strength is a "Largest is Best" measure which transforms to the following signal to noise statement:

$$S/N = -10 \log \left[1/n \sum_{i=1}^{n} 1/y_i^2 \right]$$

A combined analysis of the response graphs and ANOVA table, show that control factors C,D,E,F and G all have a significant effect on the core/rail bond strength.

Table 5.3.3 Experiment run sequence and results

TRIAL RUNS	FACTOR LABEL							RESULTS			
	A	B	C	D	E	F	G	Repl.1		Repl.2	
								Pos1	Pos2	Pos1	Pos2
1	20	5	30	30	5	4	20	3.98	5.23	0.58	0.64
2	20	5	30	60	20	4	100	7.12	2.82	8.99	6.38
3	50	15	120	30	5	4	20	3.92	3.92	6.16	5.86
4	20	15	120	30	20	12	100	8.45	3.47	8.66	2.53
5	50	5	120	60	5	12	100	10.79	10.07	10.08	9.06
6	50	15	30	60	20	12	20	2.73	0.0	0.96	0.0
7	50	15	30	30	20	4	20	6.97	0.0	3.34	0.0
8	50	5	120	60	20	4	20	3.88	3.56	6.69	2.81
9	20	15	120	60	5	4	100	11.39	10.25	12.15	12.19
10	50	5	30	30	5	12	100	8.87	0.42	9.74	7.19
11	20	15	30	60	5	12	20	0.66	0.0	1.06	0.23
12	20	5	120	30	20	12	20	1.90	0.0	1.90	0.76

Randomising sequence:
replicate 1 .... trial numbers 6 5 9 4 12 7 8 2 11 1 10 3
replicate 2 .... trial numbers 7 6 5 9 4 2 10 1 11 3 12 8

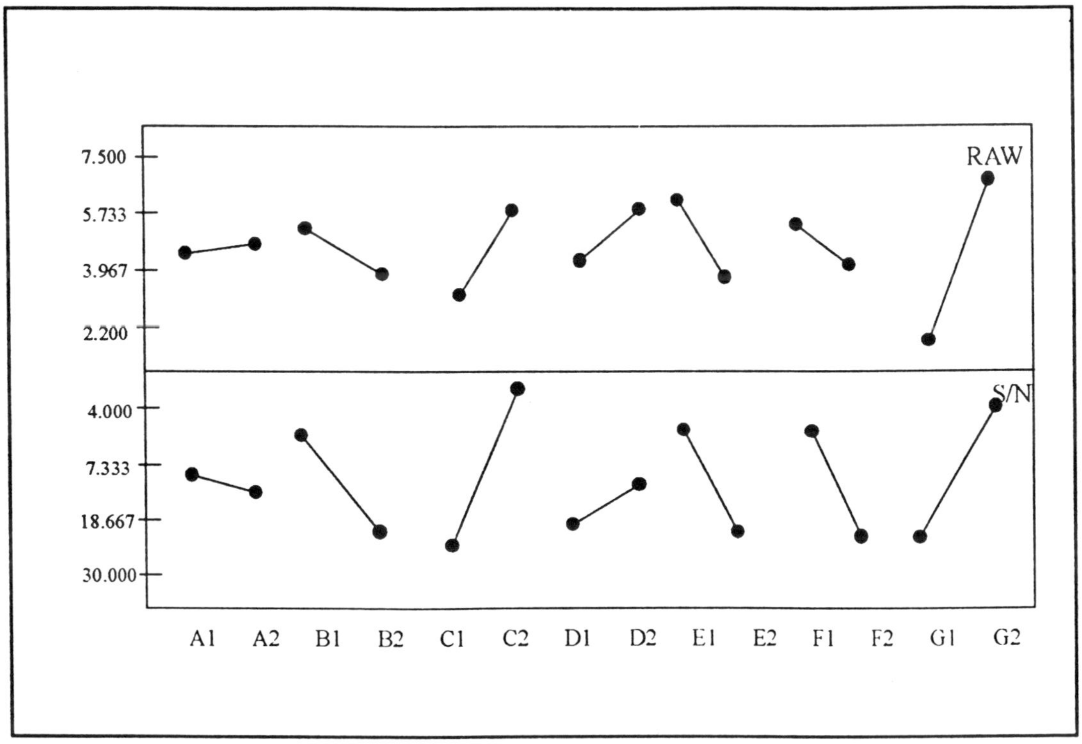

Figure 5.3.3 Response graphs

Table 5.3.4 Table of response averages at different levels

LEVEL AVERAGES (RAW DATA)							
	FACTOR						
LEVEL	A	B	C	D	E	F	G
1	4.64	5.14	3.25	3.94	6.02	5.37	2.39
2	4.88	4.37	6.27	5.58	3.5	4.15	7.12
Difference	0.24	0.77	3.02	1.64	2.52	1.22	4.73

Table 5.3.5 ANOVA Table

ANOVA TABLE (RAW DATA)					
Source	**df**	**SS**	**MS**	**F Ratio**	**Significance**
A	1	0.673	0.673	0.134	0.284
B	1	7.173	7.173	1.434	0.761
C	1	109.614	109.614	21.918	0.999 *
D	1	32.308	32.308	6.460	0.985 *
E	1	76.296	76.296	15.256	0.999 *
F	1	17.885	17.885	3.576	0.933 *
G	1	268.371	268.371	53.663	0.999 *
Other	4	28.968	7.242	1.448	0.762
Error	36	180.05	5.001 (an estimate of variance)		
Total	47	721.34			

Key: df = degrees of freedom
SS = sum of squares
MS = mean squares

5.3.13 The optimum process

From the analysis, the optimum bonding process can be defined by setting the most significant control factors at the level which produced the highest level average (see Table 5.3.4), as follows:-

C at level 2
D at level 2
E at level 1
F at level 1
G at level 2

Since factors A and B show no significance there is freedom to choose whichever level best suits the production process, hence A was set at ambient temperature (level 1) and B at 5 minutes (level 1).

It is possible to predict the bond strength from the optimum process using only the highly significant control factors, as follows:-

Predicted Bond Strengths = C2 + D2 + E1 + F1 + G2 - 4 (x bar)

where; C2, D2 etc are the level average strengths for these factors and x bar is the overall mean.

Using the above equation with the level average strengths shown in Table 5.3.4, the predicted bond strength using the optimum process was 11.327kN. The prediction can be bounded with a confidence level (see reference below) and a 90% confidence boundary in this case was $\pm$ 1.603kN.

Reference Ross PJ. Taguchi Techniques for Quality Engineering. 1988. pp 121.

5.3.14 The confirmation trial

It is always essential to confirm the results from the experiment with a confirmation trial before committing the process to permanent change. For the confirmation trial six cores (twelve core/rail bonds) were prepared using the following optimum process.

A	Parts Temperature	20 °C
B	Fixture Time	5 minutes
C	Cure Time	120 minutes
D	Oven Time	60 minutes
E	Solvent Time	5 minutes
F	Cure Time	4 hours
G	Oven Temperature	100 °C

This process resulted in a mean bond strength of 11.410kN with the lowest value being 8.900kN which was very close to the prediction. A probability plot for the confirmation trial results showed that the optimum process was very robust with an insignificant probability of producing parts that would fail the 2kN min strength requirements.

5.3.15 Implementation of the optimum process

The optimum process (defined above) was implemented at two vendors and early results of bond strength were as follows:-

Vendor 1	Mean Bond Strength	9.67kN
	minimum	7.69kN
Vendor 2	Mean Bond Strength	10.18kN
	minimum	7.70kN

It should be noted that the bond strengths achieved at the vendors were a little lower than those achieved from the laboratory experiments. However, the probability of any parts not achieving the 2kN minimum requirements was again insignificant so no further investigation was undertaken.

5.3.16 Conclusion

The DOE/ "Taguchi" methodology has been shown to be a successful method of optimising an adhesive process and, in the case of core/rail bond, more than 350,000 units (700,000 bonds) have been produced with no failures.

The structured approach is the backbone to the success by ensuring that no items are missed. Initially, a clear definition of the problem, objectives and measurement system are important to ensure correct focus to the work.

The "brainstorm" is the most crucial step because no amount of statistical theory can compensate for a critical factor that has been missed and the views of ALL the people involved in the process must be considered in the selection of the control factors for the experiment.

Finally, the confirmation trial is essential, to ratify the selection of variables to be investigated and the experimental analysis, before the process is installed in production. A committed team approach is needed to ensure that the discipline of the methodology is maintained to a successful completion, since short cuts will normally lead to failure.

There has been much interest in the "Taguchi" methodology and IBM Havant is currently supporting NPL and other industrial collaborators (PIRA, Taywood Engineering, SATRA and British Steel) in a DTI programme on enhancing adhesive technology. One particular task in this programme is to investigate the applicability of the DOE/ "Taguchi" methodology in optimising adhesive bonding processes across a number of industrial sectors including, packaging, construction, footwear and mechanical engineering.

5.3.17 Acknowledgement

This case study was a pilot for the widespread and worthwhile application of Taguchi methods at IBM Havant. Its success was due to the commitment and enthusiasm of Liz Dunn for identifying the problem and conducting the experiments.

5.4 Failure Mode and Effect Analysis (FMEA) - Law 80 Launch Tube, Hunting Engineering

Failure mode and effect analysis (FMEA) is the analysis of potential failure modes to determine their effects. The primary objective of FMEA is to determine the features of product design, production operation and distribution which are critical to the various possible modes of failure. It uses all available experience and expertise to identify the importance, levels or criticality of potential problems and simulate action at the design stage. The crucial elements of a complete analysis are:

Failure Mode — The anticipated conditions of operation are used as a base to study the most probable failure modes.

Failure Effect — The potential failures are studied to determine their possible effects on the performance of the whole product.

Failure Causes — Identifying failure mode should lead us to the possible causes of such failures.

Failure Criticality — The severity of each failure, the criticality index or risk priority number (RPN), can be assessed by multiplication of three parameters that are numerically scaled between 1 to 10. These parameters are:

P– probability of each mode occurring,
S– the seriousness or criticality of failure,
D– the difficulty of detecting the failure.

Corrective Action — Finally, a corrective action has to be suggested in order to prevent failure occurring, for the first time, or again.

Table 5.4.1 shows a generic FMEA analysis for an adhesive bonding application. It covers all common stages in an adhesive bonding application and the most common modes of failure have been displayed for each stage. It is impossible to cover all possible failure modes that may occur in an adhesive bonding application and the main objectives of this generic analysis are:

1 to provide a list of common failure modes at the various stages of the adhesive bonding process.

2 to demonstrate the logical relationship between modes of failure and causes of failure and also to show what consequences these failures may have on an adhesive bonded joint.

3 to show typical actions which could be taken to prevent such failures.

Table 5.4.1 (a) A generic FMEA for adhesive bonding

Failure Mode and Effect Analysis (FMEA)

Part Name: Part fabricated
Date: 25th Nov. 1994
Page: 1

Performed By: K.Ebtehaj
Supervisor:
Distribution: QUASIAT Sponsors

No	Process/Function 1	Possible Failure Mode 2	Effect of Failure 3	Possible Cause(s) of Failure 4	P	S	D	C	Corrective Action 7
	INWARD MATERIALS								
1	**Adhesive - Bonding**	Wrong base materials Mismatched parts Expired materials Contaminated / badly stored Wrong size or form Not up to the required strength	Weak bonding, WB/early failure No bonding, NB Weak bonding/ early failure Weak bonding/ early failure Delay in processing, DIP Weak bonding/ early failure	Wrong selection by supplier / designer Wrong despatch by supplier Wrong despatch by supplier Badly stored / contaminated area Wrong despatch by supplier Wrong selection by supplier / designer					Correct the design.Supplier QC Cert. Return to supplier/ get the correct parts Controlled storage Controlled storage & working area Return to supplier/ get the correct parts Supplier QC Cert./ change the design.
2	**Consumables (Pre-bonding) (Post-bonding)**	Wrong base agent Wrong grade / concentration Contaminated agent	No acceptable treatment/ WB Inefficient treatment/ WB Inefficient treatment/ WB	Wrong selection by supplier / designer Wrong despatch by supplier Badly stored / contaminated area					Correct the design.Supplier QC Cert. Return to supplier/ get the correct parts Controlled storage & working area
3	**Adherends**	Wrong base materials Wrong grade wrong dimension Not strong, flexible enough Wrong form & colour	Weak / No bonding Weak bonding Loose/ No fitting Early service failure Mismatched/ unacceptable	Wrong selection by supplier / designer Wrong selection by supplier / designer Incorrect processing or design Incorrect specifications by supplier Wrong despatch by supplier					Correct the design. Supplier QC Cert. Correct the design. Supplier QC Cert. Correct the design. QC on processing. Supplier QC cert. Supplier QC cert.
	PROCESSING								
4	**Prefitting**	Incorrect matching parts	Rejected product Weak bonding	Bad design, Operator error					Correct the design, use Poka-Yoke technique.
5	**Surface Treatment** **Degreasing** **Roughening** **Chemical** **Physical**	Residual contamination Residual contamination Inadequate finish Residual contamination Inadequate finish Less active surface Inadequate finish	Weak bonding Weak bonding Weak bonding Weak bonding Weak bonding Weak bonding Weak bonding	Operator error, Inadequate rinsing Operator error, Inadequate rinsing Inaccurate operation - operator error Inadequate rinsing - operator error Inaccurate operation Inaccurate operation Inaccurate operation					Train operator, Random sample testing Train operator, Random sample testing Train operator, Random sample testing Train operator, Random sample testing Use a well prepared instruction, SPC Use a well prepared instruction, SPC Use a well prepared instruction, SPC

Table 5.4.1 (b)　　A generic FMEA for adhesive bonding

Failure Mode and Effect Analysis (FMEA)

Part Name:
Date:
Page: 2

Performed By:
Supervisor:
Distribution:

No	Process/Function 1	Possible Failure Mode 2	Effect of Failure 3	Possible Cause(s) of Failure 4	P	S	D	C	Corrective Action 7
6	**Assembly** **Application** **Bonding**	Incorrect dispensing Lack of adhesion Misplaced parts Thick glue line Mismatched parts	Uneven/ Corrugated finish, WB Weak bonding Weak bonding Weak bonding No currect functioning	Applicator failure, Operator error, Using High/ low viscosity adhesive, Wrong Mixing, Contaminated adhesive, Overdue handling, Badly Fixed (Unsuitable design) Overdue dispensing, operator error					Train operator, QC on dispensing. QC on assembly schedule, Change the design, QC on Dispensing QC on fabrication (i.e. Poka-Yoke)
7	**Curing**	Under-cured Over-cured	Low durability, low strength Low strength, corrugated finish	Low temp. Short period. Uneven heat High temp. Long period. Uneven heat					SPC on Temp., Automatic switch off SPC on Temp., Automatic switch off
8	**Final Inspection**	Scattered qualification test results.	Unreliable Inspection.	Test badly conducted (operator error), Machines not calibrated.					

The specific example used to demonstrate this technique comes from the manufacturing process of Law 80, at Hunting Engineering, where nylon fittings are bonded to a composite tube, as documented in APPENDIX A. Due to unavailability of all the processing details, numerical assessment of risk priority number (RPN) is approximate. In this example two stages are defined. The first stage is the inspection of the incoming materials (adhesive, adherends and consumable materials). The second stage is the fabrication of the joint which includes surface preparation, assembly and curing.

In the first row of Table 5.4.2, the adhesive used for this application has been studied. Two possible modes of failure have been considered, 1- degraded adhesive and 2- wrong initiator. In respect to these modes of failure the initial and final effect of failure has been identified. By determining the possible cause of failure it is possible to suggest a corrective action. The final stage of analysis is to assess RPN by defining P, S and D.

This can be done by estimating a reasonable value for each of them after careful consideration of the actual manufacturing processes in relation to these parameters. When a full analysis has been carried out the Risk Priority Number (RPN) is calculated for each function or process. The RPN value represents the criticality of each function. Figure 5.4.1 shows the result of this analysis in a bar chart, where the total RPN for each function is presented, with respect to the mode(s) of failure. The highest risk of failure will be found in the assembly stage, within which 21% of the risk is due to uneven dispensing, 42% is due to excess adhesive (and hence overflow), and the rest due to misplaced joints. By controlling the assembly stages correctly, then the risk of failure can be significantly reduced. The recommended corrective actions listed in the last column of the FMEA table are easy to be applied and they are very effective. These corrective actions aim to reduce P, S, D or any combination of them, i.e. temperature control reduces P & D whereas pumping air out of the applicator mainly reduces S. The automated dispensing and part handling methods that were applied to this assembly process, to ensure controlled and reproducible bonding, have been published and are referenced in Section 6.2.

FMEA is a simple and straightforward method to help in the control of design and manufacture of an adhesively bonded product. Although it cannot display the entire interfunctional relationship which exists in a typical adhesive bonding application, the final conclusion is presented in the form of failure modes, causes and their effects. The final effects in the above example were generally limited to a few, known defects. Weak bonding was the only final effect recorded in practice, despite the process containing a range of failure modes. The important benefit of FMEA is that a corrective action will be designated to each mode of failure and this provides several, good opportunities for assembly staff to maintain a tightly controlled assembly line.

Note however, that this technique also has limitations when it is used with an adhesive bonding application. When change or adjustment is recommended to one or more features of the design or process (adhesive, adherends, joint design or processing capability), it is likely that adjustment of other features will also be required.

Table 5.4.2 (a) FMEA for Law 80 launch tube

Failure Mode and Effect Analysis (FMEA)

Part Name: Fabricated reinforced Nylon housing ring.
Date: 25th Nov. 1994
Page: 1

Performed By: K.Ebtehaj
Supervisor:
Distribution: QUASIAT Sponsors

No	Process/Function	Possible Failure Mode	Effect of Failure: Initial	Effect of Failure: Final	Possible Cause(s) of Failure	P	S	D	C	Corrective Action
	INWARD MATERIALS									
1	**Adhesive - Bonding** Toughened Acrylic + Initiator	Degraded adhesive		Weak bonding	Incorrect storage, contaminated working area.	1	7	2	14	Controlled storage, ventilated working area.
		Wrong initiator	Process failure		Incorrect despatch by supplier	1	9	2	18	Controlled inward adhesive
2	**Consumables - Pre bonding** Grit sand - 25 um Sand paper - 100	Wrong grade	Badly treated surfaces	Weak bonding	Incorrect despatch by supplier	1	2	2	4	Controlled inward consumables
	Ultrasonic cleaning agents	Wrong brand	Contaminated surface	Weak bonding	Incorrect despatch by supplier	1	3	2	4	
3	**Adherends** Reinforced Nylon	Outside tolerance	Bad fitting	Weak bonding	Badly processed - operator	2	4	4	16	Prefit control with qualified part.
	Kevlar/Epoxy Composite	Damaged coating		Weak bonding	Bad packaging, handling	1	4	3	12	Controlled packaging
	PROCESSING									
4	**Surface Treatment - Grit blasting** Grit blasting - Sanding	Over/under treated		Weak bonding	Over/under rate blasting, Short/long time of treatment.	2	2	4	16	Use of "Instruction Sheet"
	Ultrasonic Cleaning	poor surfaces		Weak bonding	Under rate vibration. Short time	1	2	3	6	Controlled rate and timing.

Table 5.4.2 (b) FMEA for Law 80 launch tube

Failure Mode and Effect Analysis (FMEA)

Part Name: | **Performed By:**
Date: | **Supervisor:**
Page: 2 | **Distribution:**

No	Process/Function	Possible Failure Mode	Effect of Failure: Initial	Effect of Failure: Final	Possible Cause(s) of Failure	P	S	D	C	Corrective Action
5	**Assembly**									
	Application	Patchy dispensing		Weak bonding	Low temperature	1	6	2	12	Temperature control.
					Blocked nozzle,	2	6	2	24	Clean nozzle after 12 runs.
		Excess adhesive	Thick bond line	Weak bonding	Trapped air,	2	6	2	24	Pump air out.
					End of container.	3	6	2	30	Use of sensor.
					High temperature	2	4	2	16	Temperature control.
					Wrong nozzle	1	4	2	8	Control nozzle before run.
	Bonding	Misplaced Joint	Poor fit	Weak bonding	Misplaced parts,	2	6	3	36	Use a fixing frame.
					fixture tools are badly placed.	2	6	2	24	Use instruction.
6	**Curing**									
		Incomplete curing	Poor handling strength.	Weak bonding	Wrong temperature,	1	2	3	6	Controlled temperature and time.
					short/ long cure period.	1	1	4	4	

P = The probability of failure (1 = low chance of occurence -to- 10 = almost certain to occur)

S = The seriousness or criticality of the failure (1 = not serious - to - 10 = total failure, safety hazard)

D = The difficulty of detecting the failure before the product is used by the customer (1 = easily detected - to - 10 = unlikely to be detected).

C = Calculate the product of ratings, **C = P x S x D,** Known as the criticality index or risk priority number (**RPN**) for each failure mode.

Table 5.4.2(c) - Cumulative RPNs

Process / Function	RPN	Total
Adhesive	14, 18	32
Adherends	16, 12	28
Surface treatment	16, 6	22
Assembly (Application and Bonding)	12, 24, 30, 16, 8, 36, 24, 24	174
Consumables	4, 4	8
Cure	6, 4	10

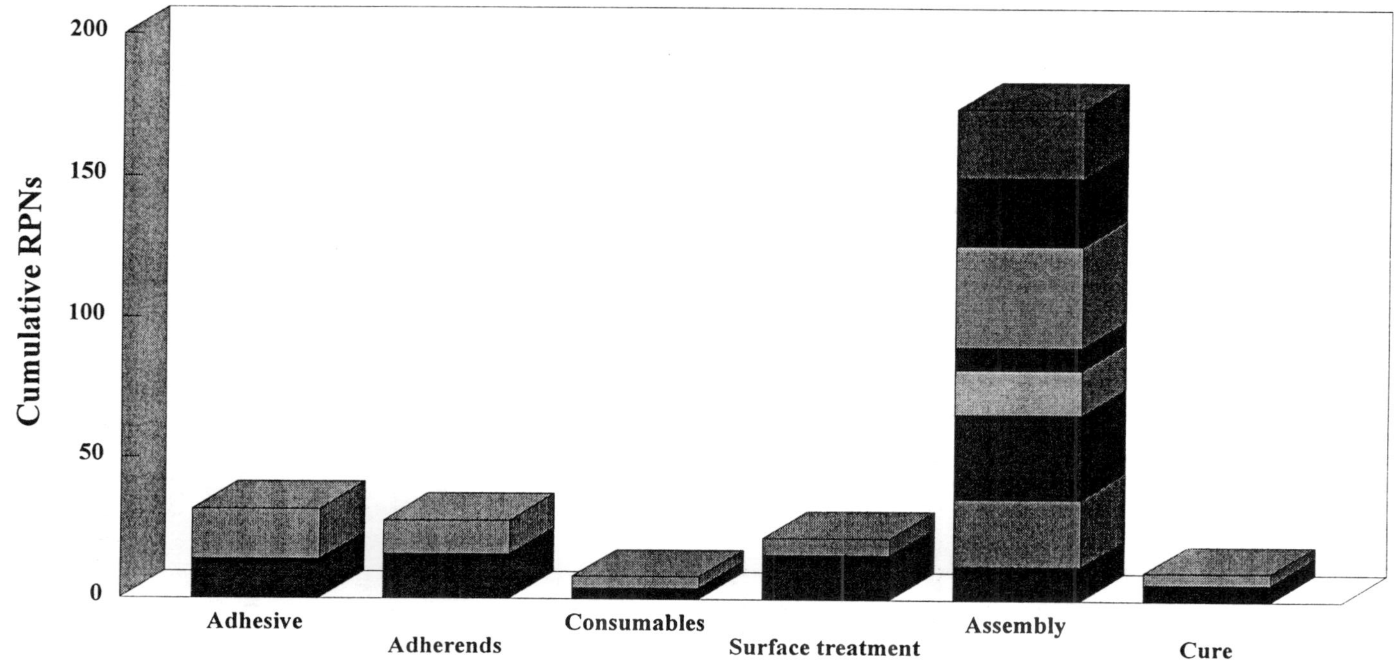

Figure 5.4.1 Cumulative RPNs from an FMEA analysis

5.5 Ishikawa Diagrams (Cause and Effect) - Power Unit Structural Framework,Commercial Hydraulics Keelavite

In a statistical process control system, statistics provide the necessary baseline information. However, when a process has gone out of control, they do not identify the cause or the remedial measures. Once a defect, error, or a problem has been identified and isolated, further study has to be carried out to analyse potential causes of the undesirable effect. Where causes are not obvious the Ishikawa Diagram (Cause and Effect, or Fishbone Diagram) is a formal and valuable tool in revealing potential causes. The steps in constructing the cause and effect diagram are as follows;

1 - define the problem or effect to be analysed
2 - form the team to perform the analysis
3 - draw the effect box and the centre line
4 - specify the major potential cause categories and join them as boxes connected to the centre line
5 - identify the possible causes and classify them into the categories in step 4. Create new categories, if it is necessary
6 - rank the causes to identify those that seem most likely to impact on the problem
7 - take an appropriate corrective action

Constructing a Cause and Effect Diagram is a team activity and often the team will uncover potential causes through brainstorming. Constructing such a diagram starts with generic causes and detailed (more specific) causes will be added to it as the analysis proceeds. Constructing a detailed diagram prior to the team involvement should be avoided since it may adversely affect team enthusiasm and lead to incorrect selection of causes. Nevertheless, a generic diagram, where the major possible causes for an application have been defined, can be a great help in bringing unknown (uncommon) causes to the surface in a brainstorming session.

In Figure 5.5.1 a generic diagram is presented for the adhesive bonding process. It consists of the five major categories of causes, Information, Materials, People, Equipment, and Procedures. Each major category is then divided into the several related causes , with details specific to adhesive bonding.

The power unit structural framework, fabricated in aluminium and aluminium honeycomb, and documented in APPENDIX A has been selected to demonstrate the application of this quality tool. Data from statistical process control measurements show that only 1800 products out of 2000 produced in ten shifts are satisfactory and conform to the design schedule. The undesirable effects in this production period are classified as follows.

43.5% are due to weak bonding
34.0% are due to rejected product
22.5% are due to delays in the process

Four causes of **weak bonding**, which may be responsible for this effect, are as follows.

1 - wrong brand of adhesive (which in turn may be due to insufficient information from the supplier or incorrect selection by the designer)

2 - incorrectly mixed adhesive (operator error)

3 - incorrect dispensing (operator error or lack of correct instruction - production manager/engineer error)

4 - use of the unsuitable post bonding cleaning agent (designer error, deterioration of agent, operator error).

Figure 5.5.2 shows the Fishbone Diagram where the possible causes of weak bonding are highlighted.

A similar approach is taken to analyse the other two effects - **rejected product** and **process delay**. Figures 5.5.3 and 5.5.4 contain the Fishbone Diagrams for these effects. Finally, a complete diagram can be constructed by combining these three diagrams into one (Figure 5.5.5).

Once the possible causes have been defined, then the weight that each cause carries should be estimated. Unrealistic or insignificant causes should be separated from those which have significant impact. Statistics and other information from the design, production, inspection and purchasing areas can help to weight each cause with more confidence. Table 5.5.1 and Figure 5.5.6 show the results of such an investigation for the current example.

When the cause has been defined, appropriate corrective actions can be taken to ensure that such errors will not be repeated. In this example, surface preparation appears to be the greatest cause of problems. The corrective actions recommended in this case are as follows.

1 - Use a standard method for surface preparation
2 - Display a step-by-step process instruction at the workshop
3 - Train the operator

The Cause and Effect Diagram can be applied to most adhesive applications, but the results are most powerful when the tool is applied to a manufacturing process which has quantitative, statistical data available (e.g. a high rate of production or a continuous process). It is one of the simplest and most effective methods to analyse statistical data gathered from a manufacturing process. The necessary steps to construct the Cause and Effect Diagram are well defined and they are easy to follow. The complex interfunctional relationships, which are characteristic of most adhesive bonding applications, are not obvious in this technique, though many are likely to become apparent during the brainstorming session.

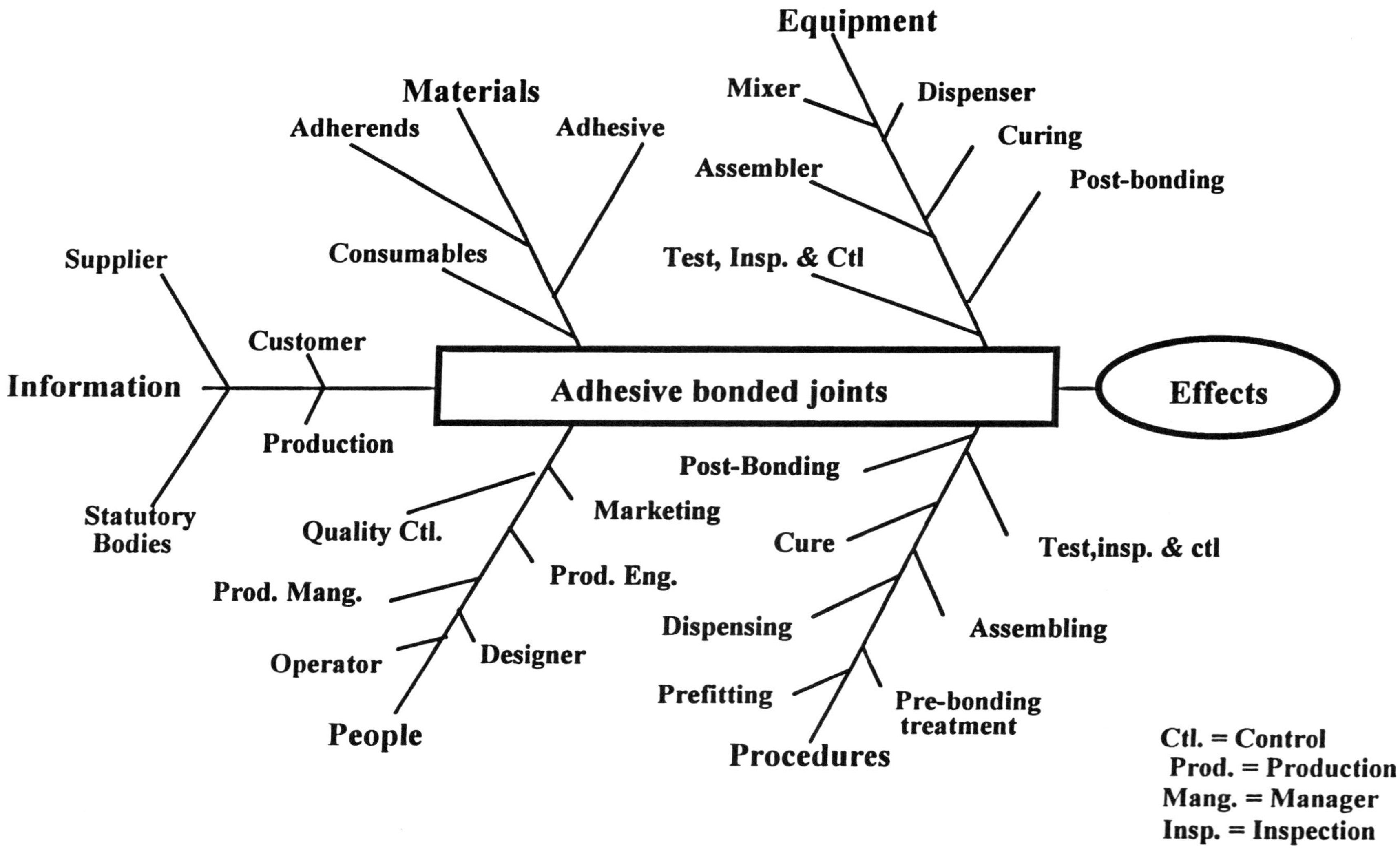

Fig. 5.5.1 A generic Fishbone Diagram for adhesive bonding applications

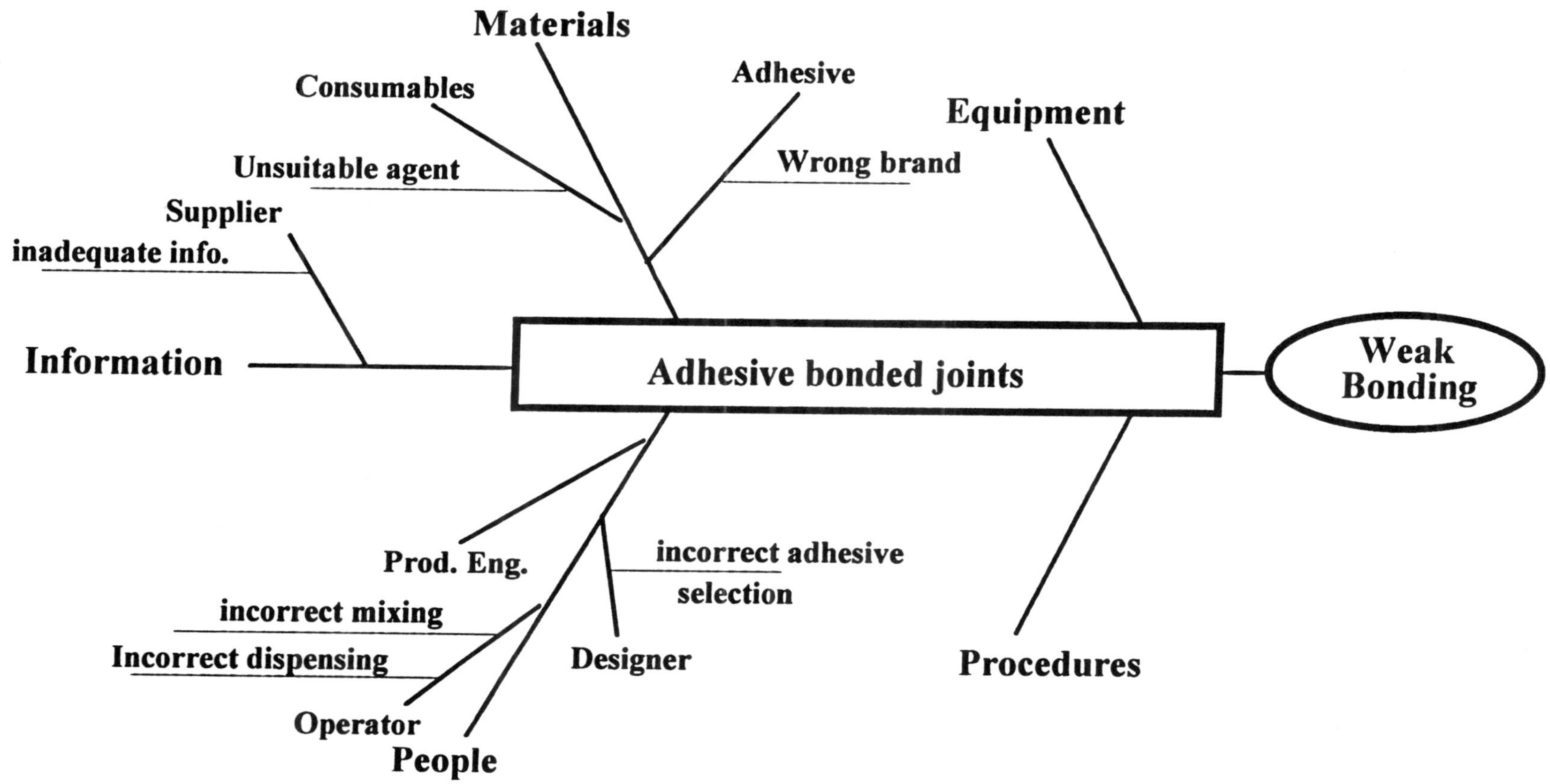

Fig. 5.5.2 Fishbone Diagram analysing the weak bonding effect

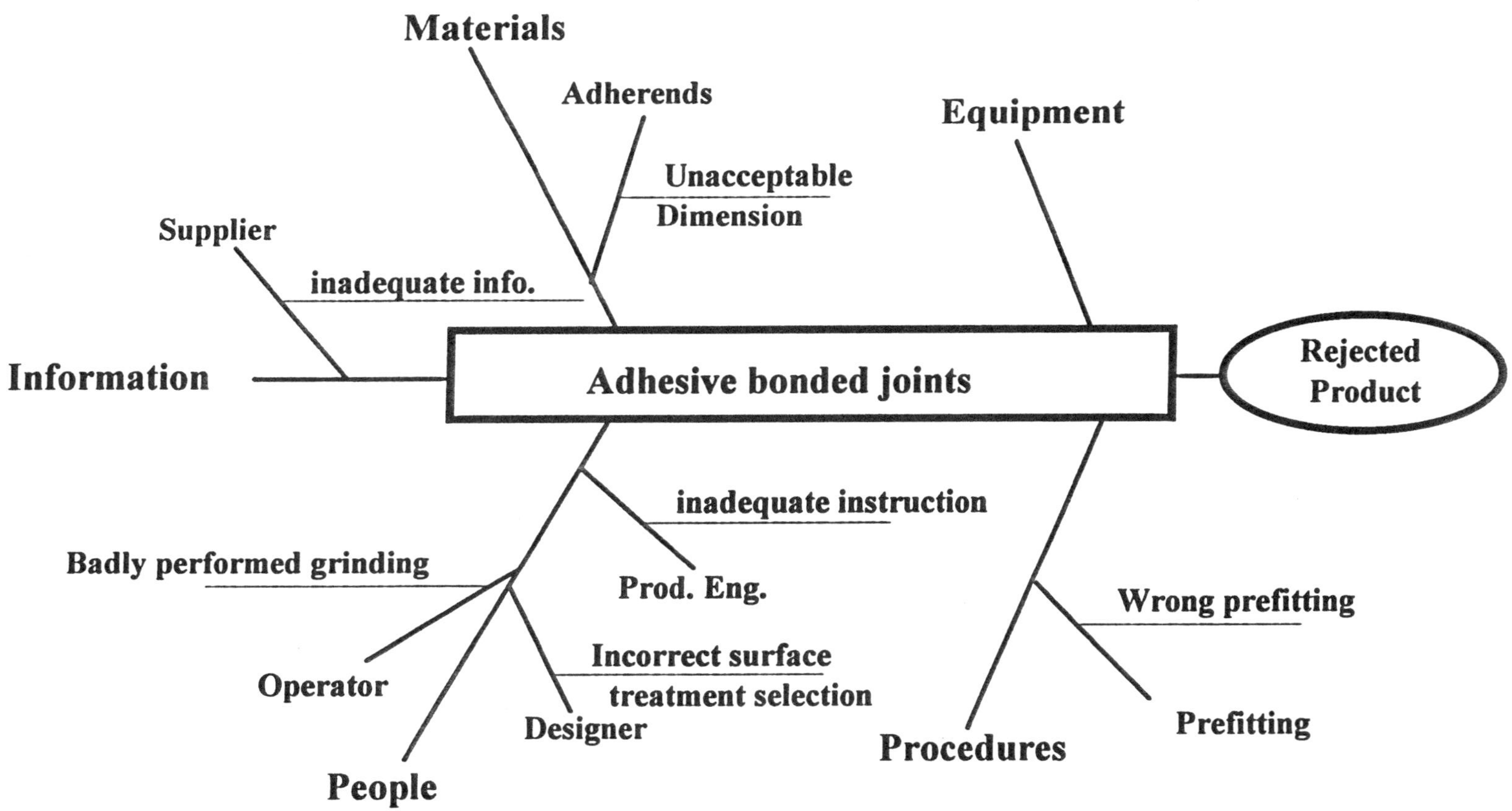

Fig. 5.5.3 Fishbone Diagram analysing the rejected product effect

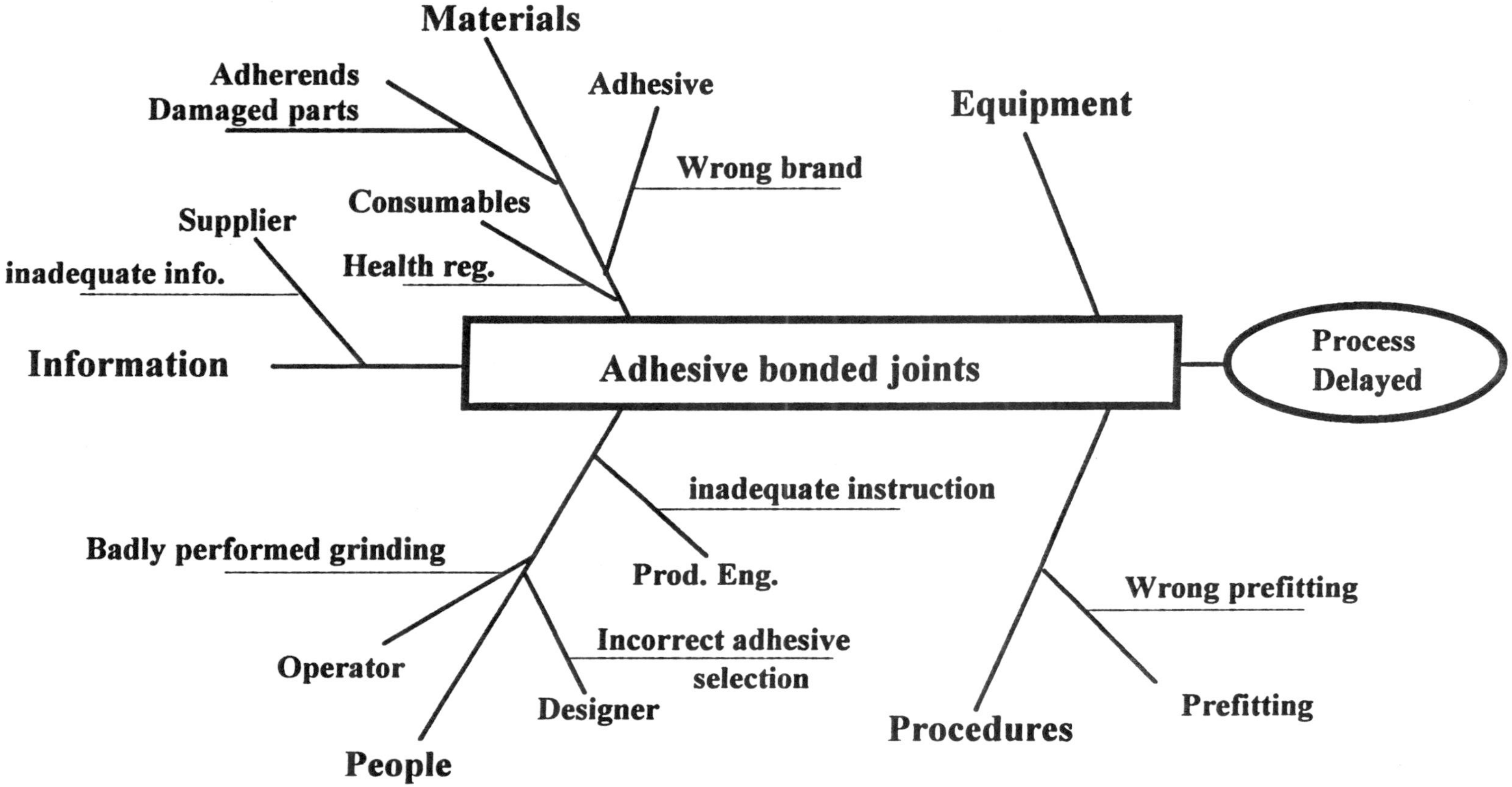

Fig. 5.5.4 Fishbone Diagram analysing the process delayed effect

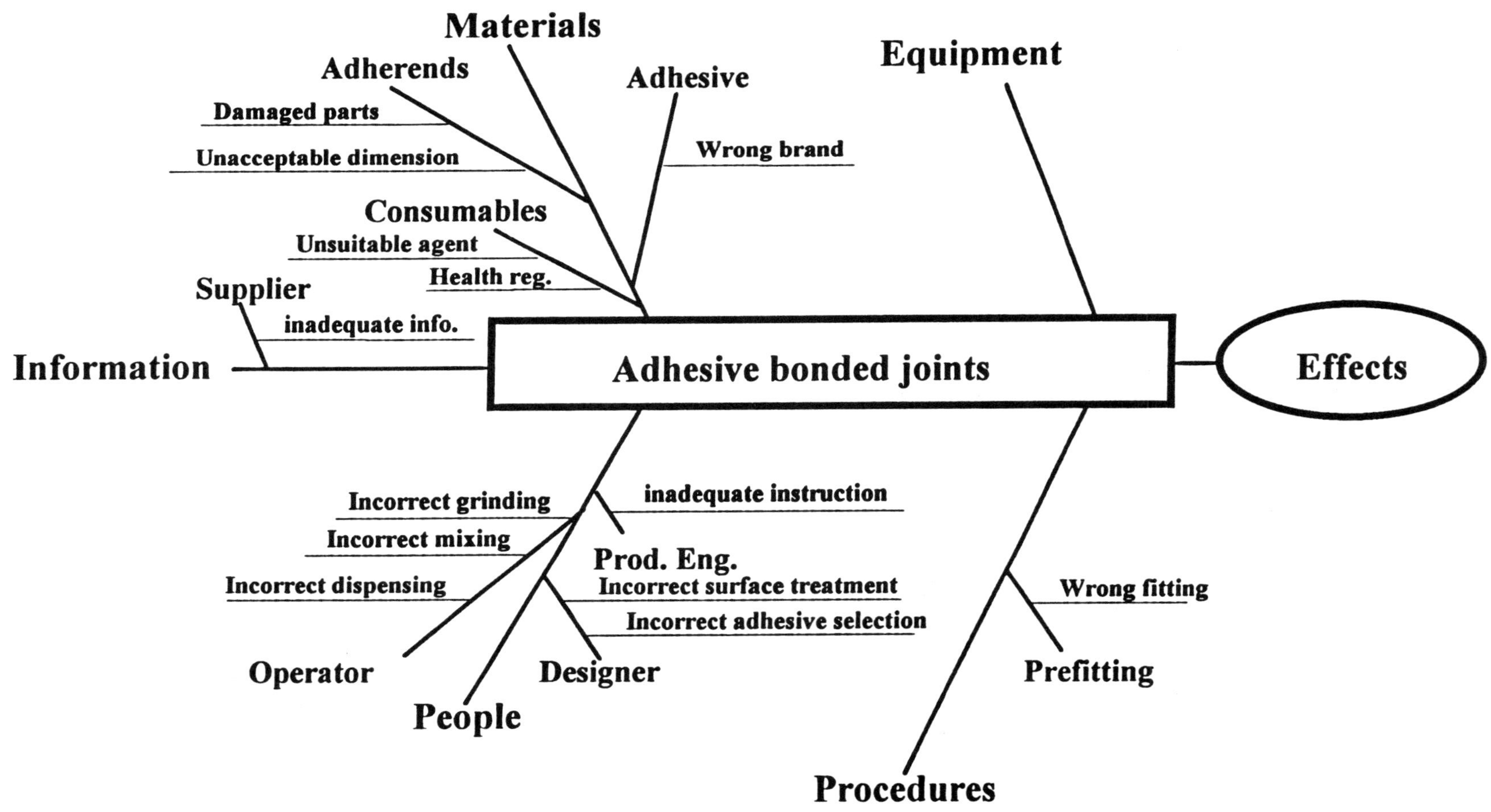

Fig. 5.5.5 Fishbone diagram, bonding of power unit structural framework

Table 5.5.1 Cause and Effect, and corrective action, bonding of power unit structural framework

Incidents	%	Effect	Corrective Action
Wrong brand of adhesive	3.0 2.0	Weak bonding Delay on the assembly schedule	QC on purshasing, QC on design
Cleaning agent hasn't satisfied health reg.	2.5	Delay on the assembly schedule	QC on design
Damaged honeycomb before delivery	8.0	Delay on the assembly schedule	QC on inspection of the incoming materials
Dimension beyond acceptable tolerance	0.5 3.0	Mainly scrapped (cost increased). Delay on the assembly schedule	QC on inspection of the incoming materials
Wrong prefitting	1.0 6.0	Mainly scrapped (cost increased). Delay on the assembly schedule.	Poke-Yoka control on prefitting Use a well prepared instruction
Badly performed surface grinding	32.5	Rejected product.	Train operator QC on design, and incoming materials
Incorrectly mixed adhesive	8.5	Weak bonding.	Train operator Use a well prepared instruction
Adhesive is dispensed incorrectly	21.0	Weak bonding.	Train operator Use a well prepared instruction
Use of the unsuitable post-bonding cleaning agent	11.0	Weak bonding.	QC on design QC on inspection of the incoming materials

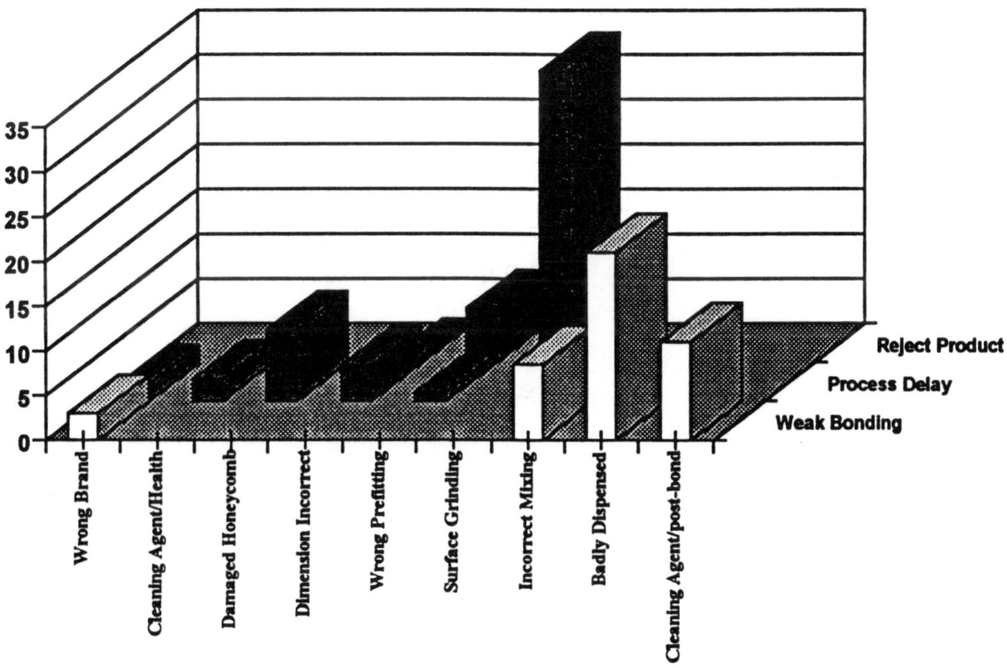

Fig. 5.5.6 Cause and Effect, and weighting, bonding of power unit structural framework

5.6 Poka-Yoke (Mistake-Proofing)

One of the prime objectives of anyone carrying out an activity is to deliver defect-free products or services. Traditional inspection of the final product is not cost efficient and is time consuming. Also this end-of -the-line method cannot ensure a 100% quality product.

Poka-Yoke systems are designed to pick up the mistakes in an operation. They prevent errors (**causes**) being converted into defective products (**effects**). They prevent defects at source, ensuring the problem is either not allowed to happen, or stop the operation to ensure that the fault is recognised before it passes down the assembly line. Two simple but very effective applications of Poka-Yoke in adhesive technology are presented in this section.

5.6.1 Bonded composite aircraft skin panel, Westland Aerospace

In the construction of lightweight composite aircraft skin panels, several layers of composite and various parts with different shapes and dimensions have to be bonded together. Between 20 to 100 parts sized from 6 to 600 cm^2 may be bonded together. Prefitting is the common practice to ensure that all parts fit together well and the designer often considers the possibility of constructing a loose assembly, prior to the final bonding step. However, mistakes made by the operator in these circumstances are not uncommon and consist of the following.

1. Incorrect components are bonded together.
2. Bonding of parts is not carried out in the correct order on the assembly line.

Ensuring that the right part will be used in the right order of assembly is an important task. To avoid the cost of reversing human mistakes, which could be expensive or even impossible in such applications, parts are identified according to their order of assembly. Also the mating parts are designed in an **asymmetric shape** so they will not fit together unless they are in the correct orientation.

Although asymmetric design can prevent bonding the wrong parts together, it does not control the order of assembly. The second device used in this application is **number coding**. Parts are numbered according to their order of assembly during prefitting, so that mistakes can be prevented during bonding. It may be argued that this device does not guarantee a mistake-proof process since the operator may still bond parts out of their correct order. However, SPC measurements used to control the bonding process show that no mistake has been made since this device was initiated eight years ago.

5.6.2 Motorcycle luggage pannier, Honda

Control of bondline thickness is important for the optimum performance of an adhesive joint and many, sometimes complex methods can be used. In the example shown below in Figure 5.6.1, taken from a Honda motorcycle luggage pannier, two parts of ABS plastic are joined and sealed by a combination of ultrasonic spot-welding and adhesive bonding.

The periodic spot-welds provide high strength attachment points with resistance to stress in all failure modes, including peel, where adhesives are weakest and they jig the parts while the epoxy cures to a load-bearing joint. The adhesive will contribute significantly to the final joint performance, plus it seals the entire joint length at a cheaper assembly cost than continuous welding.

The Poka-Yoke device is the design of joint detail in this hybrid combination. Pressure of the welding tools ensures direct part/part contact at the welded side of the joint and squeeze-out of excess epoxy from the adhesive joint. Dispensing of a slight excess of adhesive onto the part will guarantee a filled joint, always of optimum thickness, in this simple, but elegant joint design.

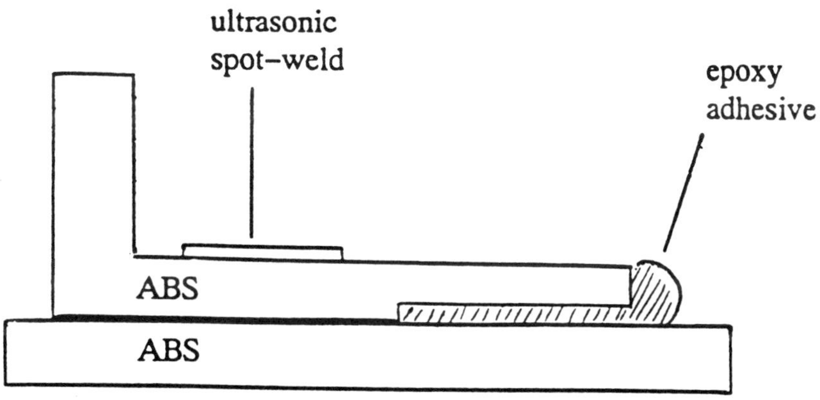

Figure 5.6.1 joint detail from Honda motorcycle luggage pannier

6 SELECTED FURTHER READING

6.1 Quality Management References

1. Tools and Techniques for Quality Management. Society of Motor Manufacturers and Traders, 1991.

2. The Quality Gurus. (DTI Booklet from 'Managing into the 90's' programme).

3. E.M. Rooney, J.H. Rogerson: Measuring Quality Related Costs. Chartered Institute of Management Accountants, 1992. ISBN 0 9 48036 90 7.

4. J.M. Juran: Quality Control Handbook. McGraw-Hill, New York, 1988.

5. G. Taguchi: Introduction to Quality Engineering. Asian Productivity Organisation, Tokyo, 1986.

6. K. Ishikawa: Guide to Quality Control (2nd edition). Asian Productivity Organisation.

7. L.E. Stebbing: Quality Assurance (3rd edition). Ellis Horwood, 1993. ISBN 0 13 334559 9.

8. J.S. Oakland: Total Quality Management.

9. P.D.T. O'Connor: Practical Reliability Engineering (2nd edition). John Wiley, 1985. ISBN 0 471 90551 8.

10. E.R. Ott: Process Quality Control - Troubleshooting and Interpretation of Data. McGraw-Hill. ISBN 0 07 047923 2.

11. P.G. Leaney and G Wittenberg: Design for Assembling, Assembly Automation, Vol 12 (2), 1992, pages 8-17.

12. A. K. V. Jones: Quality management the Nissan way, in Managing Quality, (Eds. B. G. Dale and J. J. Blunkett), Philip Allan, Hemel Hempstead, 1990, pages 44-54. ISBN 0 86003 657 X

6.2 Adhesive Technology References

1. D. M. Brewis and D. Briggs (Eds.): Industrial Adhesion Problems, Orbital Press, Oxford, 1985. ISBN 0 946193 01 0.

2. A.J. Kinloch: Adhesion and Adhesives: Science and Technology, Chapman and Hall, London, 1987.

3. W. A. Lees: Adhesives in Engineering Design, The Design Council, Springer-Verlag Publishers, London, 1984.

4. H. Brinson (Ed.): Adhesives and Sealants, ASM handbook Vol 3, ASM International, 1990. ISBN 0-87170 281 9

5. D. E. Packham (Ed.): Handbook of Adhesion, Longman Scientific and Technical, Harlow, 1992. ISBN 0 470 21870 3

6. R. Woolman and A. Hutchinson: Resealing of Buildings - a Guide to Good Practice, Butterworth Heinemann, Oxford, 1994. ISBN 0 7506 1859 0

7. Anon: Robots provide consistency in large volume adhesive bonding (Law 80), European Adhesives and Sealants, September 1991, pages 20-21.

8. R. M. Allanson: Adhesive bonding - an effective way to join aircraft structures, Welding and Metal Fabrication, October 1994, pages 384-386.

9. F. Telo and W. Knight: DFA takes a new look at Adhesives, Machine Design, January 24, 1994, pages 67-70.

APPENDIX A MANUFACTURING PROCESS CHECKLISTS

Adhesion Institute, Delft	GRP composite pipework
Anonymous	Flexible packaging lamination
British Steel Technical	Laminated floor panel
Carello Lighting	Headlamp assembly
Commercial Hydraulics Keelavite	Power unit structural framework
Hunting Engineering	Law 80 launch tube
IBM UK	Disk drive actuator
Oxford Brookes University	Construction joint sealing
Permabond	Needle electrode
Pilkington	Automotive glazing
Westland Aerospace	Aircraft wing vane
Blank Checklist	

QUASIAT, EU716			
Process plan / checklist	Company		Gelukkig - Lijm
	Site		Windmill View, Leiden, Netherlands
Centre for Adhesive Technology, TWI	Line		
	Process	development	
		production	on-site installation
DATE: 25/1/94	Product	intermediate	
FILE: pipe1.wb1			
		final	**Structural bonding of**
			composite (GRP) pipework
	Authors		K Guyt,
Blank file is PROCESS.WB1			**Adhesion Institute, Delft**
Borland Quattro Pro	Approved for release		yes
5.0 for Windows	Comments		

adhesive		
manufacturer		Ciba Polymers
grade		Wavistrong adhesive type 3 (XB5029/XB5030)
incoming specification		
supplier QA status		ISO 9002
acceptance test(s)		
sampling basis		
primary polymer type		2 part epoxy
formulation (critical factors)	fillers	note if present
	stability	3 year at room temperature if not mixed
	rheology	
storage (critical factors)	temperature	0 < T <35 °C
	atmosphere	
	packaging	250 gram kit, contains part A and B,
		sanding paper and spatula
	transport	
	shelf life	expiry date given on each kit
	rate of use	varies, depending on pipe diameter

description of part		filament-wound glass / epoxy pipe
		socket end
manufacturer		Wavin Repox BV
grade		Wavistrong
incoming specification		
supplier QA status		ISO 9001
acceptance test(s)		
sampling basis		
key attributes (critical factors)	physical form	glass / epoxy pipe
	chemical composition	fully cured epoxy resin
	mechanical properties	pressure class
shape (critical factors)	dimensions	socket of correct cone angle (e.g. 20°)
	tolerance	defined by calibration machine
surface condition (critical factors)	as received	
	known contamination	oil, grease or moisture
	existing coating	
required surface condition prior to bonding		clean, dry and free from abrasion debris
		sanded within 1 hour of bonding
pretreatment(1)	process	solvent wipe
	materials	acetone, MEK or MIBK and clean cloth
	monitoring methods	operator
pretreatment(2)	process	sanding of cone face and stop shoulder
	materials	flapper sander or sanding bobbin
		then wiped with clean, dry cloth
	monitoring methods	abraded surfaces should look clean and dull
		not excessive (or resin-rich inner layer removed)
pretreatment(3)	process	dry the wet surface
	materials	heating blanket
	monitoring methods	surface temperature must be > ambient

description of part		filament-wound glass / epoxy pipe
		spigot end
manufacturer		Wavin Repox BV
grade		Wavistrong
incoming specification		
supplier QA status		ISO 9001
acceptance test(s)		
sampling basis		
key attributes (critical factors)	physical form	glass / epoxy pipe
	chemical composition	fully cured epoxy resin
	mechanical properties	pressure class
shape (critical factors)	dimensions	spigot of correct cone angle (e.g. 20°)
	tolerance	defined by calibration machine
surface condition (critical factors)	as received	
	known contamination	oil, grease or moisture
	existing coating	
required surface condition prior to bonding		clean, dry and free from abrasion debris
		sanded or calibrated within 1 hour of bonding
pretreatment(1)	process	solvent wipe
	materials	acetone, MEK or MIBK and clean cloth
	monitoring methods	operator
pretreatment(2)	process	sanding of spigot face or calibration
	materials	flapper sander or calibrating machine
	monitoring methods	then wiped with clean, dry cloth
		abraded surfaces should look clean and dull
		not excessive (or resin-rich inner layer removed)
pretreatment(3)	process	dry the wet surface
	materials	heating blanket
	monitoring methods	surface temperature must be > ambient

process

description of process		bonding of socket / spigot joint
specification		to be carried out by company-certified staff only
adhesive application (values, methods and tolerances)	temperature	mix and apply at 20 <T < 35 °C
	metering	components supplied in pre-weighed kit
	mixing	mix all of A and B in each 250g kit
	dispensing	by hand with spatula
	monitoring method	mix for >3 min until mix has uniform colour
		use within setting time (40 min at 20 °C)
joint (critical factors)	geometry	socket and spigot coaxial joint
	bond gap	defined by cone angles in joint faces
order of assembly / layup		spread 0.5-0.8mm adhesive on inner socket face
		spread 0.8-1.0mm adhesive on outer spigot face
		insert spigot, rotate 1/4 turn (if no filler in adhesive)
		tap home with wooden block on pipe end
		remove excess adhesive
		apply heating blanket
assembly environment	temperature	site temperature, 15 < T < 40 °C
	cleanliness	
prior to cure (critical factors)	jigging / clamping	self-locating
	aggressive chemicals	none
cure (values, methods and tolerances)	method	heating blanket
	time	1 hour to heat up, 1 hour at temperature
	temperature	130 °C
	pressure	ambient
	atmosphere	air
	monitoring method	operator checks temperature and time
		blanket switch on/off times marked on joint
type	continuous	
	batch	X
	single	
	rate of production	repair or plant installation 10-20 / day
		pipeline installation >100 / day

product

inspection tests	visual	correct alignment, excess adhesive visible
		and cured (by scratch with sharp object)
	physical	liquid pressure test
	NDT	X-ray and ultrasonics (techniques in development)
	destructive	during development work and if customer wants
	sampling basis	pressure test every day during installation
	pass criteria	no leaks
performance tests		liquid pressure test , time /pressure profile
	sampling basis	one per length of one day installation
	pass criteria	at 1.5 times working pressure for 4 hours,
		then pressure drop< 0.5 bar per 1000 metres
		periodic visual inspection (if possible)

QUASIAT, EU716

Process plan / checklist	Company		**Anonymous**
	Site		
Centre for Adhesive Technology, TWI	Line		
	Process	development	
		production	example
DATE: 13/7/92	Product	intermediate	
FILE: fpl1.wb1			
		final	**Flexible packaging lamination,**
			printed PET to LDPE
Blank file is	Authors		A W Espie
PROCESS.WB1	Approved for release		approved for general release
Borland Quattro Pro	Comments		
5.0 for Windows			

adhesive

manufacturer		Morton International
grade		specific grade number, or use internal reference
incoming specification		Data sheets with recommended coat weight/cure/safety
supplier QA status		send QA results with each batch
acceptance test(s)		visual - colour, gels, separation etc.
sampling basis		each batch
primary polymer type		2-part polyurethane, 100% solids
formulation (critical factors)	fillers	no
	stability	related to storage
	rheology	
	Mix ratio	1 : 100
storage (critical factors)	temperature	<20 °C
	atmosphere	no
	packaging	size of container for catalyst and resin
	transport	yes
	shelf life	6 months

adherend(1)

description of part		polyester film
manufacturer / grade		ICI / Melinex 800
incoming specification		data sheet
supplier QA status		supply test data
acceptance test(s)		gauge (thickness), substance, visual
sampling basis		1/batch or 1 reel in 10
key attributes (critical factors)	physical form	film (12 micron thick)
	chemical composition	PET
	mechanical properties	tensile strength, burst, shrinkage
shape (critical factors)	dimensions	width, length, gauge to specification, or as required
	tolerance	+/- 10%
surface condition (critical factors)	as received	visual - no damage
	known contamination	primer, coating
	existing coating	surface treatment - note which side is print barrier
required surface condition prior to bonding		pretreated
pretreatment(s)	process	corona discharge
	materials	in line discharge unit
primer(s)	process	
	materials	print, or other coating
monitoring methods		print quality, by strobe light
notes		Surface tension
		Print - correct colours/design for customer initial sample.
		Note surface to be bonded, i.e. print may be external or sandwiched.

adherend(2)

description of part		LDPE
manufacturer / grade		Shell
incoming specification		data sheet
supplier QA status		supply test data
acceptance test(s)		gauge, substance, visual
sampling basis		1/batch or 1 in 10 reels
key attributes (critical factors)	physical form	film (50 microns thick)
	chemical composition	grade - homopolymer, copolymer or blend
	mechanical properties	tensile, burst
shape (critical factors)	dimensions	width, length, gauge to specification
	tolerance	+/- 10%
surface condition (critical factors)	as received	visual - gels, reel quality
	known contamination	coextrusion or monolayer?
	existing coating	
required surface condition prior to bonding		pretreated
pretreatment(s)	process	corona discharge - at extension stage -
	materials	in line with lamination
primer(s)	process	
	materials	
monitoring methods		No back treatment, therefore identify surface to be bonded.
		surface tension, reel quality
		uneven ribs, gels

process

description of process		flexible packaging film lamination
specification		identify machine number and type / roller setup
		job number, customer, machine settings, materials
		example of print design attached from previous run, or agreed
		feed path and pressure nip
adhesive application (values, methods and tolerances)	temperature	e.g. 20 - 90 °C
	metering	coat weight - roller type - gravure
	mixing	ratio catalyst: resin e.g. 100 : 1
	dispensing	close to feed roller
	monitoring method	coat weight at setup - top up reservoir
safety issues		gloves, goggles, cleaning after use
joint (critical factors)	geometry	pressure nip
	bond gap	temperature 100 °C
		pressure 100 units
order of assembly / layup		surface to be coated e.g. PET
assembly temperature		roller temperature - chill rollers
prior to cure (critical factors)	jigging / clamping	
	aggressive chemicals	
cure (values, methods and tolerances)	time	5 - 10 days
	temperature	20 - 50 °C, room temp. or hot oven
	pressure	
	atmosphere	
	monitoring method	chart record
type	continuous	yes
	batch	
	single	
	rate of production	100 - 300 metres / min

product

inspection tests	visual	yes
	physical	bond strength - room temp. and accelerated cure
	NDT	?
	destructive	
	sampling basis	at setup, then 1 reel / shift
	pass criteria	visual and bond strength
performance tests		
	sampling basis	after cure
	pass criteria	bond strength, ageing, chemical / product resistance

QUASIAT, EU716

Process plan / checklist Centre for Adhesive Technology, TWI DATE: 13/7/92 FILE: bst1.wb1 Blank file is PROCESS.WB1 Borland Quattro Pro 5.0 for Windows			
	Company		Anonymous
	Site		
	Line		
	Process	development	
		production	yes
	Product	intermediate	
		final	**Laminated raised-access floor panel**
	Authors		T B Jones,
			British Steel Technical, Welsh Laboratories
	Approved for release		approved for general release
	Comments		

adhesive

manufacturer / grade		Swift Adhesives
incoming specification		
supplier QA status		
acceptance test(s)		
sampling basis		
primary polymer type		moisture curing polyurethane, in solvent
formulation (critical factors)	fillers	
	stability	
	rheology	
storage (critical factors)	temperature	Room temperature
	atmosphere	exclude moisture before use
	packaging	sealed drums, 1.45 gallons
	transport	
	shelf life	

adherend(1)

description of part		Hot dip zinc coated steel
manufacturer / grade		British Steel
incoming specification		
supplier QA status		Registered to BS 5750 part 2
acceptance test(s)		None except absence of oil on surface.
sampling basis		None
key attributes (critical factors)	physical form	flat sheets and trays
	chemical composition	
	mechanical properties	To BS1449
shape (critical factors)	dimensions	From 1 metre square to 1 metre x 3 metres
	tolerance	Flatness is critical
surface condition (critical factors)	as received	untreated
	known contamination	Supplied non-oiled / handling kept to a minimum
	existing coating	Chromate passivation on surface
required surface condition prior to bonding		as received
pretreatment(s)	process	none
	materials	none
primer(s)	process	two-pack etch primer
	materials	
monitoring methods		visual examination

adherend(2)

description of part		chipboard panel
manufacturer / grade		N/A
incoming specification		must satisfy density specification
supplier QA status		unknown
acceptance test(s)		bending tests
sampling basis		
key attributes (critical factors)	physical form	Highly densely packed fibres
	chemical composition	
	mechanical properties	stiffness
shape (critical factors)	dimensions	1 metre square to 1 metre x 3 metres
	tolerance	must fit into tray
surface condition (critical factors)	as received	untreated
	known contamination	moisture
	existing coating	none
required surface condition prior to bonding		dry with no evidence of contamination
pretreatment(s)	process	none
	materials	none
primer(s)	process	none
	materials	none
monitoring methods		visual examination

process

specification		
adhesive application (values, methods and tolerances)	temperature	room temperature
	metering	automated spray equipment
	mixing	no
	dispensing	one-part
	monitoring method	visual coverage of adhesive
joint (critical factors)	geometry	rectangular
	bond gap	0.5mm - 1.0mm
order of assembly / layup		spray galvinised tray, lay in chipboard,
		spray closing plate
assembly temperature		room temperature
prior to cure (critical factors)	jigging / clamping	passage through pinch rolls
	aggressive chemicals	none
cure (values, methods and tolerances)	time	24 hours
	temperature	room temperature
	pressure	Laminate stacked into 2 metre high packs
	atmosphere	factory
	monitoring method	none
type	continuous	yes
	batch	
	single	
	rate of production	variable

product

inspection tests	visual	yes
	physical	bending test
	NDT	none
	destructive	tear-down if fails bending test, or suspect batch reported by customer
	sampling basis	variable, but very infrequent
	pass criteria	sufficient stiffness
performance tests		deflection test
	sampling basis	
	pass criteria	deflection of laminate must not exceed agreed value

QUASIAT, EU716			
Process plan / checklist	Company		**Carello Lighting PLC**
	Site		Cannock, UK
Centre for Adhesive Technology, TWI	Line		Headlamp Assembly
	Process	development	
		production	Yes
DATE: 9/9/92 FILE: carel1.wb1	Product	intermediate	Lens/Body Assembly
		final	**Automotive Headlamp**
Blank file is PROCESS.WB1 Borland Quattro Pro 5.0 for Windows	Authors		David Boucher / Alan Espie
	Approved for release		approved for general release
	Comments		

adhesive		
manufacturer		National Starch & Chemical
grade		Bondmaster H130C Hot Melt Sealant.
incoming specification		Carello Purchasing Specification CL60-2001
supplier QA status		Developing to special or preferred status as part of Carello Supplier Liaison Programme
acceptance test(s)		As per Carello Purchasing Specification CL60-2001.
sampling basis		Every batch for viscosity, filler, infra-red trace.
primary polymer type		Butyl Rubber based
formulation (critical factors)	fillers	19-23% undiscolsed minerals
	stability	12 months unopened
	rheology	5.5 - 7.5 x 10*5 mPa.s @ 175 °C
storage (critical factors)	temperature	10 - 25 °C
	atmosphere	Ambient Air
	packaging	Cardboard tubes with release paper
	transport	Conventional
	shelf life	12 months unopened.

description of part		headlamp body
manufacturer		ICI Materials
grades	Filled Polypropylene	'Procom' 1) T20C556 2) T40H550 3) GC30H251
	Filler	1) 20% Talc 2) 40% Talc 3) 30% Glass Coupled
incoming specification		1) CL40-2500 2) CL40-2501 3) CL40-2502
supplier QA status		Preferred as part of Carello Supplier Liaison Programme
acceptance test(s)		To CL40-2500 or 2501 or 2502
sampling basis		Random; Every 10-20 Batches for MFI, filler, polymer type
key attributes (critical factors)	physical form	Injection Moulded Component
	chemical composition	Filled polypropylene.
	mechanical properties	Tensile Yield, elongation, Flexural Modulus.
		Notched Impact. Heat Distortion Temperature.
shape (critical factors)	dimensions	Body Channel (recommendations in design guide)
	tolerance	
surface condition (critical factors)	as received	Injection moulded with no pre-treatment.
	known contamination	Possible by external mould release agents , also general dust
	existing coating	None
required surface condition prior to bonding		Not specified
pretreatment(s)	process	None
	materials	None
primer(s)	process	None
	materials	None
monitoring methods		None

description of part		headlamp lens
manufacturer		Holophane and Vetreric Riunite
grade		soda lime glass
incoming specification		Lucas. 197-5-1 and 197-5-2
supplier QA status		Developing to limited or preferred status as part of Carello Supplier Liaison Programme.
acceptance test(s)		Lucas 197-5-3 and developed visual assessment methods
sampling basis		Occasional for thermal shock test.
key attributes (critical factors)	physical form	Moulded Glass Lens
	chemical composition	Propriety
	mechanical properties	Thermal Shock
shape (critical factors)	dimensions	Lens Location foot (recommendations in design guide)
	tolerance	
surface condition (critical factors)	as received	No treatment
	known contamination	Suspected that Release Agents are used
	existing coating	None
required surface condition prior to bonding		Surface Energy > 56 mN/m
pretreatment(s)	process	None
	materials	None
primer(s)	process	None
	materials	None
monitoring methods		surface tension ink

process

description of process		automatic robot or pantographic dispensing
specification		Assembly operation specified as part of
		Manufacturing Process plan.
adhesive application (values, methods and tolerances)	temperature	Dispensing temperature of melt 160-180 °C
	metering	None
	mixing	None
	dispensing	By gear pump - speed critical
	monitoring method	Visual, and SPC - by weight of 5 consecutive
		shots taken twice per 8 hrs shift
joint (critical factors)	geometry	Lens to Body (recommendations in design guide)
	bond gap	
order of assembly / layup		Sealant dispensed into Body Channel, lens
		placed manually or robotically within 15 seconds
assembly temperature		Not known accurately
prior to cure (critical factors)	jigging / clamping	Clamp immediately after lens placement
	aggressive chemicals	None
cure (values, methods and tolerances)	time	Five minutes allowed before pressure test
	temperature	Melt dispensed at 160-180 °C cool down before lens fitment
	pressure	Atmospheric
	atmosphere	Air
	monitoring method	Controlled by line speed set by balance timing
type	continuous	cell-based assembly line operation carried
	batch	out on batch builds of
	single	product types.
	rate of production	1-2/min, dependant on the product and line speed

product

inspection tests	visual	yes
	physical	Air or water leak testing
	NDT	Air or water leak testing
	destructive	In the event of specific problems.
	sampling basis	!00% assembly line testing
	pass criteria	No leaks, satisfactory bond strength.
performance tests		
	sampling basis	In process testing to customer schedule.
	pass criteria	Customer Product Specification.

QUASIAT, EU716			
Process plan / checklist			
	Company		**Commercial Hydraulics Keelavite Ltd (CHKL)**
Centre for Adhesive Technology, TWI	Site		Warwick
	Line		N/A
	Process	development	
		production	yes
DATE: 22/7/92	Product	intermediate	
FILE: chkl1.wb1		final	**structural framework for defense vehicle hydraulic power pack**
Blank file is PROCESS.WB1 Borland Quattro Pro 5.0 for Windows	Authors		S Dowling
	Approved for release		approved for general release
	Comments		

adhesive		
manufacturer		Ciba Polymers
grade		Redux 403
incoming specification		Certificate of Conformance (C of C)
supplier QA status		
acceptance test(s)		None, but (C of C)
sampling basis		None
primary polymer type		Epoxy, 2-part
formulation (critical factors)	fillers	Not known
	stability	Not known
	rheology	Not known
storage (critical factors)	temperature	to be controlled
	atmosphere	Not known
	packaging	Not known
	transport	Not known
	shelf life	Controlled

adherend(1)

description of part		Aluminium Alloy BS 1470
manufacturer / grade		/ Grade 6082
incoming specification		BS 1470
supplier QA status		N/A
acceptance test(s)		None, but C of C
sampling basis		N/A
key attributes (critical factors)	physical form	
	chemical composition	aluminium alloy
	mechanical properties	
shape (critical factors)	dimensions	
	tolerance	
surface condition (critical factors)	as received	
	known contamination	
	existing coating	
required surface condition prior to bonding		Abraded, clean
pretreatment(s)	process	Abrasion, cleaning
	materials	
primer(s)	process	None
	materials	
monitoring methods		Operator

adherend(2)

description of part		aluminium skin honeycomb panel
manufacturer / grade		Grade 5251 (skin)
incoming specification		BS 1470
supplier QA status		N/A
acceptance test(s)		None, but (C of C)
sampling basis		
key attributes (critical factors)	physical form	
	chemical composition	aluminium alloy
	mechanical properties	
shape (critical factors)	dimensions	
	tolerance	
surface condition (critical factors)	as received	
	known contamination	
	existing coating	
required surface condition prior to bonding		Abraded, clean
pretreatment(s)	process	Abrade, clean
	materials	
primer(s)	process	None
	materials	
monitoring methods		Operator

process

specification		defined in WPS 020
adhesive application (values, methods and tolerances)	temperature	Room temperature
	metering	
	mixing	as instructions (ratio 100 : 7.5)
	dispensing	By hand
	monitoring method	Operator
joint (critical factors)	geometry	As drawing
	bond gap	controlled by Pop Rivets
order of assembly / layup		
assembly temperature		Room temperature
prior to cure (critical factors)	jigging / clamping	with Pop Rivets
	aggressive chemicals	N/A
cure (values, methods and tolerances)	time	16 hurs
	temperature	45 °C, +/- 5 °C
	pressure	Atmospheric
	atmosphere	Air
	monitoring method	Thermocouples, see WPS 017
type	continuous	
	batch	Occasionally 2
	single	yes
	rate of production	2 - 10 per month

product

inspection tests	visual	yes
	physical	Proof Load of Assembly
	NDT	None
	destructive	lap shear test of coupon samples
	sampling basis	defined in WPS 021
	pass criteria	N/A
performance tests		
	sampling basis	
	pass criteria	

QUASIAT, EU716			
Process plan / checklist	Company		**Hunting Engineering Ltd**
	Site		Ampthill
Centre for Adhesive Technology, TWI	Line		Law 80
	Process	development	
		production	yes
DATE: 9/9/92	Product	intermediate	**Housing ring on Kevlar / epoxy tube**
FILE: hunt2.wb1		final	**Law 80 anti-tank weapon**
Blank file is PROCESS.WB1 Borland Quattro Pro 5.0 for Windows	Authors		K D Stapleton, P R Simpson
	Approved for release		Approved for general release
	Comments		Adhesive data also supplied by
			N Pickett, Permabond

adhesive / Hunting

manufacturer / grade		Permabond F241 / MPP 360
incoming specification		AFS 1646 / DTD 5577
supplier QA status		HEL approved Grade A2
acceptance test(s)		Lap Shear Tests
sampling basis		Each supplier batch
primary polymer type		toughened acrylic
formulation (critical factors)	fillers	
	stability	
	rheology	
storage (critical factors)	temperature	+5 to +10 °C
	atmosphere	Sealed containers
	packaging	polyethylene lined steel drums
	transport	road
	shelf life	6 months

adhesive / Permabond

manufacturer / grade		Permabond F241 + MPP 360
incoming specification		AFS 1646A + HEL Curing Rate Test
supplier QA status		ISO 9001
acceptance test(s)		
sampling basis		
primary polymer type		Acrylic
formulation (critical factors)	fillers	None
	stability	Correct Mix Ratio
	rheology	High viscosity (17000-25000 m Pa.s)
storage (critical factors)	temperature	5 - 15 °C (Avoid large variations)
	atmosphere	Humidity low - medium
	packaging	Sealed, non permeable
	transport	
	shelf life	1 year from despatch. Correct stock rotation.
	Rate of use	Will relate to production rate.

adherend(1)

description of part		Injection moulded housing ring
manufacturer / grade		Scorpion Ltd
incoming specification		HEL drawing
supplier QA status		HEL approved Grade B
acceptance test(s)		
sampling basis		BS 6001 Level S3
key attributes (critical factors)	physical form	injection moulding
	chemical composition	DuPont ZYTEL ST 801
	mechanical properties	
shape (critical factors)	dimensions	Diameters
	tolerance	0.1 mm
surface condition (critical factors)	as received	as moulded
	known contamination	none
	existing coating	none
required surface condition prior to bonding		
pretreatment(s)	process	water treatment, hinging, grooving
	materials	grit blasting, ultrasonic cleaning
primer(s)	process	none
	materials	none
monitoring methods		operator control

adherend(2)

description of part		Kevlar / Epoxy Filament-wound Composite tube, coated with IR-reflective polyurethane paint
manufacturer / grade		HEL produced
incoming specification		HEL drawing
supplier QA status		NATO AQUAP 1 & 13
acceptance test(s)		Materials Process NDT Dimensional visual
sampling basis		100% inspection
key attributes (critical factors)	physical form	
	chemical composition	Epoxide Cure (Tg)
	mechanical properties	Tube Burst Strength
shape (critical factors)	dimensions	Diameters Straightness
	tolerance	0.04 mm 0.02 mm
surface condition (critical factors)	as received	as processed
	known contamination	None
	existing coating	Specified Polyurethane Paint
required surface condition prior to bonding		
pretreatment(s)	process	abraid
	materials	none
primer(s)	process	none
	materials	none
monitoring methods		Reflectivity test. Operator inspection.

process

specification		Process specification
adhesive application (values, methods and tolerances)	temperature	15 - 28 °C
	metering	Automatic metering
	mixing	static mixing head
	dispensing	robotic
	monitoring method	operator control
joint (critical factors)	geometry	squareness
	bond gap	max 0.6 mm
order of assembly / layup		N/A
assembly temperature		15 - 28 °C
prior to cure (critical factors)	jigging / clamping	Tooled fixtures
	aggressive chemicals	None
cure (values, methods and tolerances)	time	20 min handling; 3 hr test; 24 hr full.
	temperature	15 - 28 °C
	pressure	atmospheric
	atmosphere	Workshop ambient
	monitoring method	Operator or process control
type	continuous	yes
	batch	
	single	
	rate of production	180 units / day

product

inspection tests	visual	glue beads around joint
	physical	gauging
	NDT	
	destructive	shear load test
	sampling basis	1 per shift
	pass criteria	18 KN minimum
performance tests		Weapon proof test firing
	sampling basis	2000 Batch size Double 15 & 15 sampling
	pass criteria	No failure

QUASIAT, EU716			
Process plan / checklist	Company		**IBM UK Ltd**
	Site		Havant
Centre for Adhesive Technology, TWI	Line		Redwing
	Process	development	
		production	yes
DATE: 28/1/94 FILE: ibm1.wb1	Product	intermediate	core / rail assembly
		final	**PC hard disk drive actuator**
Blank file is PROCESS.WB1 Borland Quattro Pro 5.0 for Windows	Authors		M. Hall
	Approved for release		yes
	Comments		

adhesive		
manufacturer		Loctite
grade		326
incoming specification		to manufacturers specification
supplier QA status		Company approved supplier
acceptance test(s)		none
sampling basis		none
primary polymer type		modified methacrylate ester
formulation (critical factors)	fillers	not known / not tested
	stability	not known / not tested
	rheology	not known / not tested
storage (critical factors)	temperature	+5 to +28 ° C
	atmosphere	
	packaging	in manufacturers original containers
	transport	not known
	shelf life	1 year
	rate of use	

description of part		sintered core (P/N 95X2101)
manufacturer		Ames, Spain
grade		05-194 / CMH 6-2580-002 / 41-027
incoming specification		as per drawing for P/N 95X2101
supplier QA status		approved
acceptance test(s)		to dimension and surface finish requirements on drawing P/N 95X2101
sampling basis		to quality plan
key attributes (critical factors)	physical form	sintered iron / resin impregnation
	chemical composition	nickel plated
	mechanical properties	
shape (critical factors)	dimensions	groove diameter 8.04 mm
	tolerance	diameter +0.03 mm
		perpendicular to datum A < 0.05mm deviation
surface condition (critical factors)	as received	
	known contamination	protective light oil film
	existing coating	nickel plating as per std 41-027
required surface condition prior to bonding		clean and dry
pretreatment(s)	process	solvent clean prior to adhesive application
	materials	1,1,1 trichloroethane
	monitoring methods	
primer(s)	process	none applied
	materials	
	monitoring methods	

description of part		guide shaft (P/N 95X2248)
manufacturer		Kyocera, Japan
grade		sintered zirconia to IBM eng spec 95X1937
incoming specification		as per drawing for P/N 95X2248
supplier QA status		approved
acceptance test(s)		dimensions and surface roughness
sampling basis		to quality plan
key attributes (critical factors)	physical form	sintered ceramic
	chemical composition	zirconia
	mechanical properties	
shape (critical factors)	dimensions	diameter 8 mm
	tolerance	-0.02 mm
surface condition (critical factors)	as received	N3 surface roughness
	known contamination	none
	existing coating	none
required surface condition prior to bonding		clean and dry
pretreatment(s)	process	solvent cleaning prior to adhesive application
	materials	1,1,1 trichloroethane
	monitoring methods	
primer(s)	process	none applied
	materials	
	monitoring methods	

process

description of process		bonding guide shaft to core
specification		Ames manufacturing process procedure
adhesive application (values, methods and tolerances)	temperature	room temperature
	metering	N/A
	mixing	N/A
	dispensing	semi-automatic
	monitoring method	operator
joint (critical factors)	geometry	defined in assembly drawing P/N 95X1737
	bond gap	controlled by adherend geometries
		and assembly fixturing
		tolerances ensure finite gap, not interference fit
order of assembly / layup		adherends solvent cleaned
		adhesive applied to both parts
		slide parts together
		fixture for partial cure
		solvent clean excess adhesive
		oven cure (no fixture)
assembly environment	temperature	room temperature
	cleanliness	class 10,000 i.e. less demanding than
		final assembly , which is class 100
prior to cure (critical factors)	jigging / clamping	jigging geometry, spring force 100 +/- 10 N
	aggressive chemicals	1,1,1trichloroethane solvent clean after partial
		room temp. cure to remove excess adhesive
cure (values, methods and tolerances)	method	[1] tack (in fixture) [2] oven
	time	15 min 60 min
	temperature	room 100 ° C
	pressure	100 +/- 10 N N/A
	atmosphere	air air
	monitoring method	
type	continuous	
	batch	yes
	single	
	rate of production	5000 / week

product

inspection tests	visual	shaft guides free from adhesive (95X1737 note 7)
	physical	to meet drawing dimensions (P/N 95X1737)
	NDT	proof load test 2000N (95X1737 note 6)
	destructive	test to failure
	sampling basis	by statistical test plan
	pass criteria	assembly to meet NDT proof load
		audit test to failure vs. development test values
performance tests	sampling basis	N/A
	pass criteria	N/A

QUASIAT, EU716		
Process plan / checklist		
	Company	Acme Building Co.
	Site	New Street, Birmingham
Centre for Adhesive Technology, TWI		
	Process	**Joint sealing of new building**
DATE: 21/1/94	Sealant Applicator	Sealant Techniques Ltd,
FILE: seal1.wb1	Applicator status	ASA member
	Author	A Hutchinson,
Blank file is PROCESS.WB1		**Oxford Brookes University**
Borland Quattro Pro 5.0 for Windows	Approved for release	yes
	Comments	

adhesive		
manufacturer		Adshead Ratcliffe
grade		Arbokol 2000, black
incoming specification		BS 3712 , ISO 11600
supplier QA status		BS 5750
acceptance test(s)		trials (tensile and peel adhesion, staining) at start
		(important part of acceptance)
sampling basis		batch of sealant to one job
primary polymer type		2-part polysulphide,
		based on Thiokol LP Polysulphide Liquid Polymer
formulation (critical factors)	fillers	60% undisclosed mineral fillers
	stability	9 months unmixed
	rheology	part A, base 10^6 mPas
		part B, catalyst 10^4 mPas
	MAF	30% movement accomodation factor
	service life	25 years
storage (critical factors)	temperature	not exceeding 25 °C
	atmosphere	dry
	packaging	2 litre can containing part A and B
	transport	conventional
	shelf life	9 months unmixed
	rate of use	5 litres per man day
back-up material		PE foam strip, 35mm diameter
masking tape		25 mm wide

adherend(1)

description of part		precast concrete wall panels
manufacturer		Acme
grade		
incoming specification		to architect's drawing, and BS 5628 pt8, 'Use of Masonry'
supplier QA status		
acceptance test(s)		inspection prior to sealing
sampling basis		every panel
key attributes (critical factors)	physical form	flat, porous
	chemical composition	Portland cement/gravel aggregate
	mechanical properties	rigid, brittle
	coeff. of expansion	12*10^-6 /° C
shape (critical factors)	dimensions	3*2 metres
	tolerance	joint gap >> tolerance
surface condition (critical factors)	as received	laitance (cement rich surface) 2-3 mm
	known contamination	damp, dirt and loose cement dust
	existing coating	concrete surface treatments to use or remove
required surface condition prior to bonding		dry, free of loose cement or gravel and primed
pretreatment(s)	process	brush clean and solvent wipe
	materials	wire brush no.1 and Arbo Cleaner No. 13
	monitoring methods	visual, operator inspection
primer(s)	process	apply masking tape, brush on primer
	materials	Arbo Primer 925
	monitoring methods	ready when dry and seal within 6 hours

description of part		u-PVC window frame
manufacturer		Anglian
grade		
incoming specification		Anglian product spec.
supplier QA status		BS 5750
acceptance test(s)		factory audit
sampling basis		at start of contract
key attributes (critical factors)	physical form	flat profile edge
	chemical composition	u-PVC
	mechanical properties	semi-rigid
	coeff. of expansion	50*10^-6 /° C
shape (critical factors)	dimensions	2* 1 metres
	tolerance	joint gap >> tolerance
surface condition (critical factors)	as received	polyethylene masking film
	known contamination	
	existing coating	
required surface condition prior to bonding		dry and free from loose dirt
pretreatment(s)	process	remove masking film
	materials	brush clean if necessary, MEK solvent wipe
	monitoring methods	visual, operator inspection
primer(s)	process	none
	materials	
	monitoring methods	

process

description of process		sealing joint between concrete
		and u-PVC window frames
specification		Acme spec.
adhesive application (values, methods and tolerances)	temperature	surface 5-30 °C
	metering	all of B added to A
	mixing	low-speed paddle mixer, until no streaks
	dispensing	2 litre gun
	monitoring method	visual, by operator
		> 2 hours at 20 degrees C
joint (critical factors)	geometry	butt joint
	bond gap	20 - 50 mm
order of assembly / layup		clean joint faces
		place backup material
		apply masking tape
		apply primer
		apply sealant
		tool and finish
		remove tape
assembly environment	temperature	5 to 50 degrees C
	cleanliness	no rain or dust
prior to cure (critical factors)	jigging / clamping	none
	aggressive chemicals	none
	required finish	flat surface, flush with panels
cure (values, methods and tolerances)	method	
	time	7 days for 50%, 50 days for full cure
	temperature	5 to 50 °C
	pressure	
	atmosphere	
	monitoring method	hardness and interface peel at final inspection
		and interface peel
type	continuous	
	batch	X
	single	
	rate of production	30 metres / man day
daily records		weather conditions, surface temperature
		position on building (N,S,E,or W and floor)
		sealant batch no.
		sign off work completed

product

inspection tests	visual	tooled with no voids
	physical	qualitative tests
	NDT	
	destructive	sample removed only in dispute
	sampling basis	random, on behalf of client in presence of applicato
	pass criteria	adhesion to both surfaces, by finger peel
		rubber-like hardness of sealant
performance tests	sampling basis	building fabric inspection every 3-5 years
	pass criteria	adhesion to both surfaces, by finger peel,
		flexibility of sealant,
		no cracking of sealant surface,
		removal of sample and reseal, in dispute

QUASIAT, EU716			
Process plan / checklist	Company		Anonymous
	Site		
Centre for Adhesive Technology, TWI	Line		
	Process	development	
		production	yes
DATE: 27/1/94 FILE: perm2.wb1	Product	intermediate	
		final	**Needle electrode**
	Authors		N Pickett
Blank file is PROCESS.WB1 Borland Quattro Pro 5.0 for Windows			**Permabond**
	Approved for release		yes
	Comments		Permabond customer

adhesive		
manufacturer		Permabond
grade		A118 anaerobic sealant
incoming specification		Material to have viscosity 300-350 mPa.s
supplier QA status		ISO 9001
acceptance test(s)		Verification of type and quantity
sampling basis		None
primary polymer type		Acrylic (methacrylate based)
formulation (critical factors)	fillers	N/A
	stability	room temp. stable in presence of st. steel/Pt
	rheology	As stated above
storage (critical factors)	temperature	5 - 10 °C
	atmosphere	avoid excessively high humidity
	packaging	Not critical
	transport	Not critical
	shelf life	1 year at 5-25 °C
	rate of use	40 litres p.a.

adherend(1)

description of part		hypodermic syringe needle
manufacturer		Unknown - purchased via subcontractor
grade		Austenitic stainless steel
		304515 of BS 970 : 1983
incoming specification		Certificate of conformance
supplier QA status		BS 5750 Part 2
acceptance test(s)		Dimensional checks
sampling basis		To BS 6001 : Part 1 : 1972
		General inspection Level II 0.65% AQL
key attributes (critical factors)	physical form	Length and external diameter
	chemical composition	None
	mechanical properties	None
shape (critical factors)	dimensions	Straight & circular
	tolerance	No dimensions or tolerance
surface condition (critical factors)	as received	Bright
	known contamination	Grease/dirt
	existing coating	N/A
required surface condition prior to bonding		Clean, grease & dirt-free, sterile
pretreatment(s)	process	Washing of surfaces
	materials	Detergent / water
	monitoring methods	
primer(s)	process	None
	materials	
	monitoring methods	

description of part		Platinum wire
manufacturer		Johnson Matthey
grade		Platinum 390 hard Dralon
incoming specification		To user specs C80280 80281 80282
supplier QA status		BS 5750 : Part 2
acceptance test(s)		None
sampling basis		None
key attributes (critical factors)	physical form	Diameter & pin holes in coating
	chemical composition	None
	mechanical properties	No tangles
shape (critical factors)	dimensions	None
	tolerance	
surface condition (critical factors)	as received	Less than 25 pin hole/100mm
	known contamination	
	existing coating	Polyether-imide coated
required surface condition prior to bonding		
pretreatment(s)	process	None
	materials	
	monitoring methods	
primer(s)	process	None
	materials	
	monitoring methods	

process

description of process		Sealing/bonding of Pt electrode inside
		stainless steel hypodermic needles
specification		Approval to USA medical applications
adhesive application (values, methods and tolerances)	temperature	20-25 °C
	metering	Dip tray technique - 0.5" depth, 10m/s
	mixing	Single-part, N/A
	dispensing	N/A
	monitoring method	
joint (critical factors)	geometry	Concentric/coaxial
	bond gap	Small enough to allow capillary
		action to completely fill gap
order of assembly / layup		
assembly environment	temperature	20-25 °C
	cleanliness	
prior to cure (critical factors)	jigging / clamping	Needles jigged vertically
	aggressive chemicals	N/A
cure (values, methods and tolerances)	method	Oven cure
	time	20 min
	temperature	100 - 105 °C
	pressure	Normal
	atmosphere	Dry
	monitoring method	
type	continuous	
	batch	yes
	single	
	rate of production	10^6 / year

product

inspection tests	visual	cursory check based on BS6001
	physical	
	NDT	electrical continuity for shorts & opens
	destructive	sectioning & microscopy to detect voids
	sampling basis	electrical - 100%, destructive - 1 per week
	pass criteria	electrical - no shorts or opens
performance tests	sampling basis	
	pass criteria	
		USA medical approval

QUASIAT, EU716			
Process plan / checklist	Company		**Pilkington**
	Site		Eccleston, St Helens
Centre for Adhesive Technology, TWI	Line		Cup-on Line
	Process	development	
		production	Yes
DATE: 5/11/92 FILE: pilk1/wb1	Product	intermediate	**Automotive Glazing**
		final	
Blank file is PROCESS.WB1 Borland Quattro Pro 5.0 for Windows	Authors		D Pennington
	Approved for release		Approved for general release.
	Comments		

adhesive		
manufacturer / grade		National Starch reactive hot melt adhesive
incoming specification		Manufacturers specification
supplier QA status		None
acceptance test(s)		None
sampling basis		None
primary polymer type		moisture-curing polyurethane
formulation (critical factors)	fillers	
	stability	
	rheology	
storage (critical factors)	temperature	10 - 25 °C
	atmosphere	
	packaging	Tube
	transport	Conventional
	shelf life	6 months unopened

adherend(1)

description of part		toughened automotive glass
manufacturer / grade		Pilkington
incoming specification		N/A
supplier QA status		
acceptance test(s)		Clean glass
sampling basis		Every piece
key attributes (critical factors)	physical form	Pressed toughened glass
	chemical composition	Silica
	mechanical properties	
shape (critical factors)	dimensions	Smooth surface
	tolerance	
surface condition (critical factors)	as received	Clean
	known contamination	None
	existing coating	None
required surface condition prior to bonding		Clean. No contamination from toughening process
pretreatment(s)	process	None
	materials	None
primer(s)	process	wiped on via felt applicator
	materials	Sika 205 (Active Wipe)
monitoring methods		Visual inspection

adherend(2)

description of part		metal clip
manufacturer / grade		R A Labone
incoming specification		
supplier QA status		
acceptance test(s)		None
sampling basis		None
key attributes (critical factors)	physical form	Plate with pin
	chemical composition	Zinc passivated steel
	mechanical properties	
shape (critical factors)	dimensions	
	tolerance	
surface condition (critical factors)	as received	Clean
	known contamination	
	existing coating	
required surface condition prior to bonding		Clean
pretreatment(s)	process	None
	materials	None
primer(s)	process	None
	materials	None
monitoring methods		Visual inspection

process

specification		
adhesive application (values, methods and tolerances)	temperature	100 - 110 °C
	metering	Compressed air / timer
	mixing	None
	dispensing	Hot melt gun
	monitoring method	Visual
joint (critical factors)	geometry	
	bond gap	
order of assembly / layup		Primer / hot melt adhesive / stud
assembly temperature		Room temperature
prior to cure (critical factors)	jigging / clamping	Jigging
	aggressive chemicals	None
cure (values, methods and tolerances)	time	set 10 secs, final cure 3 days
	temperature	Cools to room temperature
	pressure	
	atmosphere	Atmospheric
	monitoring method	None
type	continuous	
	batch	Yes
	single	
	rate of production	1 / minute

product

inspection tests	visual	Yes
	physical	Location check
	NDT	No
	destructive	No
	sampling basis	One per eight hours
	pass criteria	Fits jig
performance tests		
	sampling basis	
	pass criteria	Pull force >20kgf (normally 40kgf)

QUASIAT, EU716			
Process plan / checklist	Company		**Westland Aerospace Ltd (WAL)**
	Site		East Cowes
Centre for Adhesive Technology, TWI	Line		MD 11
	Process	development	
		production	yes
DATE: 13/7/92	Product	intermediate	
FILE: west2.wb1		final	**Aircraft wing vane**
	Authors		H A Coulam
Blank file is PROCESS.WB1	Approved for release		approved for general release
Borland Quattro Pro			
5.0 for Windows	Comments		

adhesive		
manufacturer / grade		3M Co. AF 163
incoming specification		Customer Material Specification
supplier QA status		Approved Product / source
acceptance test(s)		Film Weight. Shear test
sampling basis		Per manufacturing batch.
primary polymer type		modified epoxy
formulation (critical factors)	fillers	carrier supported film.
	stability	fully formulated (limited life)
	rheology	essentially fixed
storage (critical factors)	temperature	-18 °C
	atmosphere	
	packaging	Moisture proof
	transport	refrigerated
	shelf life	120 days at -18 °C, 10 days RT

adherend(1)

description of part		pre-cured composite skin
manufacturer / grade		WAL (in house)
incoming specification		customer / WAL material specification
supplier QA status		WAL are approved manufacturer
acceptance test(s)		C- Scan
sampling basis		each skin
key attributes (critical factors)	physical form	aerofoil skin (5 m x lm)
	chemical composition	carbon / epoxy
	mechanical properties	care in handling / routing
shape (critical factors)	dimensions	effectively controlled by
	tolerance	curing mould & rout jug
surface condition (critical factors)	as received	with "peel ply" intact -
	known contamination	removed immediately
	existing coating	prior to bonding
required surface condition prior to bonding		abrade & solvent wipe
pretreatment(s)	process	abrade solvent wipe
	materials	wet/dry genklene
primer(s)	process	
	materials	
monitoring methods		visual inspection

adherend(2)

description of part		aluminium rib
manufacturer / grade		machined in house
incoming specification		material to non-US source
supplier QA status		(not approved)
acceptance test(s)		yes. Tensile
sampling basis		Raw material batch
key attributes (critical factors)	physical form	machined component
	chemical composition	
	mechanical properties	
shape (critical factors)	dimensions	CAD/CNC generated form
	tolerance	effects flight surface quality
surface condition (critical factors)	as received	machined surface
	known contamination	temporary protective
	existing coating	
required surface condition prior to bonding		primed
pretreatment(s)	process	degrease, alkaline clean, pickle and anodise
	materials	(approved line)
primer(s)	process	spray & air & oven dry
	materials	BR127 (corrosion resistant primer)
monitoring methods		thickness control visually primed surface - limited life

process

specification		Douglas Process Specification
adhesive application (values, methods and tolerances)	temperature	Room temperature
	metering	
	mixing	film adhesive
	dispensing	as required
	monitoring method	
joint (critical factors)	geometry	composite surface is "bag side" :. rougher than "tool side"
	bond gap	1 or 2 plies film
order of assembly / layup		skin applied to rib/adhesive assembly
assembly temperature		ambient (some use of heat gun for increased tack)
prior to cure (critical factors)	jigging / clamping	can be fit problems
	aggressive chemicals	
cure (values, methods and tolerances)	time	2 hours
	temperature	125 °C
	pressure	27 PSI
	atmosphere	non-inert
	monitoring method	part-temperatures & Vacuum Bag integrity, continuously.
type	continuous	
	batch	
	single	yes
	rate of production	10 per week

product

inspection tests	visual	visual & alignment
	physical	
	NDT	A- scan
	destructive	cured test pieces
	sampling basis	per item
	pass criteria	customer specification
performance tests		N/A
	sampling basis	
	pass criteria	

QUASIAT, EU716			
Process plan / checklist	Company		
	Site		
Centre for Adhesive Technology, TWI	Line		
	Process	development	
		production	
DATE:	Product	intermediate	
FILE:			
		final	
Blank file is	Authors		
PROCESS.WB1	Approved for release		
Borland Quattro Pro	Comments		
5.0 for Windows			

adhesive		
manufacturer		
grade		
incoming specification		
supplier QA status		
acceptance test(s)		
sampling basis		
primary polymer type		
formulation (critical factors)	fillers	
	stability	
	rheology	
storage (critical factors)	temperature	
	atmosphere	
	packaging	
	transport	
	shelf life	
	rate of use	

adherend(1)

description of part		
manufacturer		
grade		
incoming specification		
supplier QA status		
acceptance test(s)		
sampling basis		
key attributes (critical factors)	physical form	
	chemical composition	
	mechanical properties	
shape (critical factors)	dimensions	
	tolerance	
surface condition (critical factors)	as received	
	known contamination	
	existing coating	
required surface condition prior to bonding		
pretreatment(1)	process	
	materials	
	monitoring methods	
pretreatment(2)	process	
	materials	
	monitoring methods	

description of part		
manufacturer		
grade		
incoming specification		
supplier QA status		
acceptance test(s)		
sampling basis		
key attributes (critical factors)	physical form	
	chemical composition	
	mechanical properties	
shape (critical factors)	dimensions	
	tolerance	
surface condition (critical factors)	as received	
	known contamination	
	existing coating	
required surface condition prior to bonding		
pretreatment(1)	process	
	materials	
	monitoring methods	
pretreatment(2)	process	
	materials	
	monitoring methods	

description of process		
specification		
adhesive application (values, methods and tolerances)	temperature	
	metering	
	mixing	
	dispensing	
	monitoring method	
joint (critical factors)	geometry	
	bond gap	
order of assembly / layup		
assembly environment	temperature	
	cleanliness	
prior to cure (critical factors)	jigging / clamping	
	aggressive chemicals	
cure (values, methods and tolerances)	method	
	time	
	temperature	
	pressure	
	atmosphere	
	monitoring method	
type	continuous	
	batch	
	single	
	rate of production	

product		
inspection tests	visual	
	physical	
	NDT	
	destructive	
	sampling basis	
	pass criteria	
performance tests		
	sampling basis	
	pass criteria	

APPENDIX B CHECKLIST DOCUMENTS

Specification review checklists

Quality requirements

Concerned parties and data resources

Specification change

Design checklists

Design

Adherend materials

Consumable materials

Adhesive

Adhesive storage

Design of process

Specification Review Checklists

Quality Requirements Checklist

Requirement	Area	Item	Detail	Tick Here
Performance Requirements	Tolerance/Accuracy	Dimension	General Dimension	
			Bonding Area	
			Bonding Gap	
		Appearance	Shape, Size	
			Finish	
			Colour	
		Structural	Strength	
			Durability	
			Ductility	
			Shear Modulus	
		Environment		
		Cost		
		Specific Requirements		
	Quality Control	Methods		
		Records		
		Inspection		
		Control		
		Equipment		
Specification of Manufacturing Conditions	Facilities & Equipment	Prefitting		
		Pre-treatment of surfaces		
		Adhesive Mixing & Dispensing		
		Assembling, Bonding & Jigging		
		Curing		
		Storage & Despatch		
	Skill			
Safety Requirements	Compatible with	Working Atmosphere		
		Fire Regulations		
		Hygiene		
		Disposals		
Environmental/ Statutory Requirements	Inflammability			
	Toxicity			
	Odour			
	Biological Influences			

Concerned parties & Data Resources Checklist

Category	Source	Item	Tick Here
Records	**Previous data from**	**-Production Line**	
		-End User	
		-Service	
		-Storage & Despatch	
		-Materials	
		Purchasing Section	
		-Design Change	
International&National Standards		**-ASTM, BS**	
		-Manufacturing Standards	
		-Materials Supplier Standards	
		-Advance Research & Development	
		-Storage & Despatch Standards	
		-Health & Safety Regulations	
Test Work			

Specification Change Checklist

Target of the specification change is to produce a new set of specifications that conform with;

1. Customer requirements
2. Supplier specifications
3. Health & Safety regulations
4. Test works & Standards
5. Recent developments in design & manufacturing
6. Manufacturing conditions
7. Performance requirements

Design Checklists

Design Checklist

		Tick Here
Loading Conditions	Type (i.e. Cyclic, impact, ...)	
	Amplitude	
	Form (i.e. Shear, Cleavage, tensile,...)	
Integrity Requirements	Strength, Shear Modulus,...	
	Ductility	
	Durability	
Bonding Features	Bonding Gap	
	Bonding Area	
	Number & Sequence of assembly	
	Fitting (male-female) Parts	
	Surface Condition of Bonding Area	
Service Conditions	Exposure Sequence	
	Temperature & pressure Range	
	Weather, Humidity, Water, Chemical,..	
	Sunlight & Radiation	
	Outdoor (location, exposed / sheltered)	
	Indoor (ambient/controlled atmosphere)	
	Biological Influences, etc..	
Adherends	Details of Materials to be bonded. (Re: II-1- Adherend Materials Checklist)	
Consumable Materials	Details of Consumable Materials to be used. (Re: II-2- Consumable Checklist)	
Adhesives	Details of adhesives to be used for bonding. (Re: II-3- Adhesive Checklist)	
Process Conditions (Re: Design of Process Checklist)	Prefitting	
	Surface Preparation	
	Adhesive Application	
	Adhesive Assembly	
	Post-assembly Process, Cleaning, Painting, etc.	
	Curing	
	Final Inspection	
	Handling, Storage and Despatch	
	Production Rate & Preferred Condition	
	Automation, Limitation of Processing Equipment	
Quality Control & Inspection	Control/ Inspection Method (instructions)	
	Records: Method & Data acquisition	
	Equipment: Limitation, Accuracy & Service	
Service & Maintenance	Spare Parts	
	Ease of Service & Maintenance	
	Regular Inspection: Method & Instruction	

Adherend Materials Checklist

Materials	Adherends	
Properties	First	Second
Trade name		
Supplier		
Composition		
Appearance		
Form		
Surface Energy		
General Dimension		
Bonding Dimension		
Compatibility		
Static Strength		
Shear Modulus		
Ductility		
Weight		
Exposure		
Resistance to Heat		
Conductivity		
Handling		
Storage		
Despatch		
Others		

Consumable Materials Checklist

Materials	Pre-Bonding Treatments				Post-bonding Treatments
Properties	**Cleaning**	**Mechanical**	**Chemical / Electrochemical**	**Primer**	**Solvent Cleaning/ Cooling**
Trade Name					
Supplier					
Composition					
Concentration					
Form & Grade					
Storage					
Disposal					
Handling					
Others					

Adhesive Checklist

Adhesive Properties and Type		Chemical Reactive	Evaporation or Diffusion	Hot Melt	Delayed-Tack	Film Adhesive	Pressure Sensitive
Trade Name							
Applicability							
Physical Form							
Physical Properties	Viscosity Strength Stability Flexure Special Gravity Colour						
Electrical Properties	Dielectric Constant Dissipation Factor Volume Resistivity Dielectric Strength						
General Properties	Thermal Expansion Thermal Conductivity Flame Retardant Water Absorption						
Application Method							
Cure & Setting Temp. Time							
Adhesive Coverage							
Oil Tolerance							
Gap Filling							
Open Time							
Handling Time							
Functional Time							
Shelf Life							
Method of Transfer							
Bonding Pressure							
Storage Condition							
Safety Consideration							
Specific Requirements							

Adhesive Storage Checklist

Type of Adhesive	
Trade Name	
Supplier	
Batch No.	
Size of Container	
Shelf Life	
Date In	
Expire Date	
Date Out	
Storage Temperature	
Storage Humidity	
Transfer Method	
Special Requirements	

Design of Process Checklist

		Tick here
Prefitting	Method, Number of parts to be Assembled , Order of Assembly & Parts' ID (control).	
Surface Preparation	Method, Size of Adherends, Area to be treated, Required Agents, Equipment, Skilled Operator, Process Rate, Service/Maintenance, Disposal of Agents , Quality Control	
Adhesive Application	Viscosity, Tack Time, Pot-life, Open/Handle/Function Time of adhesive, Method of Application, Rate of Use Equipment/Machinery, Service, Disposal of excessive adhesive, Quality Control.	
Adhesive Assembly	Number of Assembly, Order of Assembly & Parts' ID, Bonding Pressure & fixture, Quality Control.	
Curing	Method, Volume of job, Time / Temperature of Setting and Curing, Heat Distribution Control.	
Post-assembly Processing	Cleaning, Removing excessive adhesive, Grinding/Finishing, Painting, Moulding.	
Final Inspection	Method, Equipment/Machinery, Operator/Instruction, Service/Maintenance, Control.	
Storage, Packaging & Despatch	Indoor/Outdoor Storage, Exposure to Humidity-Water-Chemical- biochemical, Number of Part per Package, External Pressure, Package Filling & Protection, Transportation.	